DEVELOPMENT TOURISM

New Directions in Tourism Analysis

Series Editor: Dimitri Ioannides, Missouri State University, USA

Although tourism is becoming increasingly popular as both a taught subject and an area for empirical investigation, the theoretical underpinnings of many approaches have tended to be eclectic and somewhat underdeveloped. However, recent developments indicate that the field of tourism studies is beginning to develop in a more theoretically informed manner, but this has not yet been matched by current publications.

The aim of this series is to fill this gap with high quality monographs or edited collections that seek to develop tourism analysis at both theoretical and substantive levels using approaches which are broadly derived from allied social science disciplines such as Sociology, Social Anthropology, Human and Social Geography, and Cultural Studies. As tourism studies covers a wide range of activities and sub fields, certain areas such as Hospitality Management and Business, which are already well provided for, would be excluded. The series will therefore fill a gap in the current overall pattern of publication.

Suggested themes to be covered by the series, either singly or in combination, include—consumption; cultural change; development; gender; globalisation; political economy; social theory; sustainability.

Also in the series

Sex Tourism in Africa
Kenya's Booming Industry
Wanjohi Kibicho
ISBN 978-0-7546-7460-3

Cultures of Mass Tourism
Doing the Mediterranean in the Age of Banal Mobilities
Edited by Pau Obrador Pons, Mike Crang and Penny Travlou
ISBN 978-0-7546-7213-5

The Framed World
Tourism, Tourists and Photography
Edited by Mike Robinson and David Picard
ISBN 978-0-7546-7368-2

Brand New Ireland?
Tourism, Development and National Identity in the Irish Republic
Michael Clancy
ISBN 978-0-7546-7631-7

Development Tourism
Lessons from Cuba

ROCHELLE SPENCER
Macquarie University, Australia

Routledge
Taylor & Francis Group

LONDON AND NEW YORK

First published 2010 by Ashgate Publishing

Published 2016 by Routledge
2 Park Square, Milton Park, Abingdon, Oxfordshire OX14 4RN
711 Third Avenue, New York, NY 10017, USA

First issued in paperback 2016

Routledge is an imprint of the Taylor & Francis Group, an informa business

British Library Cataloguing in Publication Data
Spencer, Rochelle.
    Development tourism : lessons from Cuba. -- (New directions
    in tourism analysis)
    1. Tourism--Cuba. 2. Tourism--Government policy--Cuba.
    3. Cuba--Economic conditions--1990- 4. Tourism--Social
    aspects--Cuba. 5. Cuba--Social conditions--21st century.
    I. Title II. Series
    338.4'7917291-dc22

Library of Congress Cataloging-in-Publication Data
Spencer, Rochelle.
    Development tourism : lessons from Cuba / by Rochelle Spencer.
        p. cm. -- (New directions in tourism analysis)
    Includes bibliographical references and index.
    ISBN 978-0-7546-7542-6 (hardback) -- ISBN 978-1-4094-0208-4
(ebook) 1. Tourism--Cuba. 2. Culture and tourism--Cuba. 3. Culture and globalization--
Cuba. I. Title.
    G155.C9S74 2009
    338.4'7917291--dc22

                                                                        2009038946

ISBN 13: 978-1-138-25025-3 (pbk)
ISBN 13: 978-0-7546-7542-6 (hbk)

Contents

Dedicated to Kerry Spencer and Anthony Spencer

List of Figures

Maps of Cuba

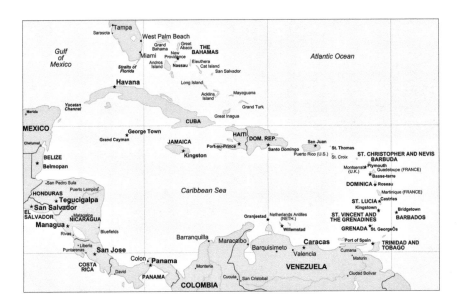

Map 1 Cuba in geographical location
Source: Adapted from Dreamstime <www.dreamstime.com>.

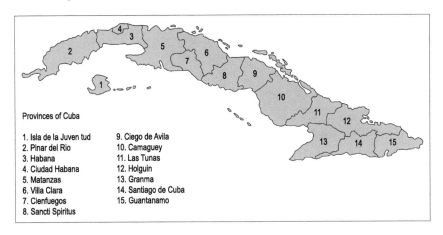

Map 2 Provinces of Cuba
Source: Adapted from Dreamstime <www.dreamstime.com>.

Maps of Cuba

Map 1 Cuba in geographical location

Source: Adapted from Thinglink, www.thinglink.com

Provinces of Cuba

1 Isla de la Juventud	9 Ciego de Ávila
2 Pinar del Río	10 Camagüey
3 Habana	11 Las Tunas
4 Ciudad Habana	12 Holguín
5 Matanzas	13 Granma
6 Villa Clara	14 Santiago de Cuba
7 Cienfuegos	15 Guantánamo
8 Sancti Spíritus	

Map 2 Provinces of Cuba

Source: Adapted from Freemanpedia, www.freemanpedia.com

Acknowledgements

I am deeply grateful to Carmen and Aida for their warm generosity and invaluable assistance in introducing me to Cuban life. They were so very helpful in assisting me to appreciate many of the issues that were crucial to my understanding of Cuba's social development. They provided me with a family environment and made my transition to life in Cuba far easier than it could otherwise have been. I am indebted to both Carmen and Aida for nursing me back to health when I was ill with Dengue Fever, or as many Cubans refer to it, Bone Crushing Disease.

My colleagues in the Centre for Research on Social Inclusion at Macquarie University provided me with great support in writing this book. Thanks especially to Rosemary Wiss and Lorraine Gibson, fellow anthropologists, for their words of wisdom.

Many thanks also to Global Exchange's *Reality Tours* for facilitating my research in Cuba. I am deeply indebted to Brian Witty and Margaret Rohde from the not-for-profit tour operator, formerly known as *Oxfam Community Aid Abroad Tours*, for giving me the once in a lifetime opportunity to work in Cuba. Over four years, this experience has had an incredible impact on me. I negotiated my way through the oppressive Cuban bureaucracy while often marveling at aspects of Cuban socialism – paradoxes abound in Cuba.

Finally, thanks to all those who shared their thoughts and impressions in the many conversations I had in Cuba, and particularly those who were interviewed for this book. Without their contributions this book could not exist.

Prologue

"We turn our setbacks into victories" – Cuban President Fidel Castro

I awake to the sound of the house keeper, Frida, rummaging noisily in the kitchen. As I wander sleepily from my bedroom I find her cleaning two dusty oil lamps. She tells me somewhat distractedly that Hurricane Michelle is approaching Cuba. Frida's anxious preparations for the hurricane make me wonder about what unknown lies ahead. I've never experienced a hurricane before. Outside the house I wonder at how calm people seem, just getting on with their day-to-day activities. I head down to the train station in Habana Vieja to purchase a ticket to Viñales, where I plan to travel with two girlfriends in the coming days. When I express concern about the approaching hurricane to the woman in the ticket office, she airily dismisses this with, 'no te preocupes', don't worry. I'm lulled into a false sense of security and begin to think that perhaps Frida is overreacting.

However, the next day it's clear that the country is in disaster preparation mode. My landlady, Sofía, explains that we need to ready the house. She frequently monitors the storm on television, each time explaining that the hurricane is advancing and strengthening. She tells me the last hurricane this big was in 1944, 2,000 people died. Sofía explains so many died because of inadequate warnings. But this time, there is constant coverage of the hurricane as it advances. President Castro appears on live television broadcasts from Cuba's weather centre, instructing and re-assuring his people continuously; I take some comfort in his reassurances.

On day three the usual clamour and clatter of busy Havana streets is distinctly absent. Normally I awake to the distinct low gear rumble and vibration of the diesel *camellos* as they make their way up *Calle San Lázaro*. But today, the atmosphere is grey and still, punctuated by the sound of hammering. Neighbourhood dogs scamper around visibly on edge. A collective sense of fear is building. People are preparing their houses, hammering shutters closed and reinforcing windows with masking tape in an effort to prevent shattering during the storm. I phone my friend who is boarding with a Cuban academic from the *Universidad de la Habana*. She tells me about her household's preparations in using wire to tie the shutters closed and further securing them to heavy wooden furniture in the apartment. I head down to the supermarket to buy some food for several days and find queues of people frantically trying to purchase food and supplies for the coming disaster. I feel a rush of dread as people push past me, clambering to grab whatever they can from the shelves. Suddenly, I feel acutely aware of the impending hurricane and I start to worry. The growing hysteria in the supermarket frenzy is confronting. Until this point I had no idea what to expect. The previous day, I had been blithely unaware of the magnitude of the coming storm and had been naively reassured by the woman who dismissed my concerns at the rail station. But now observing people in my neighbourhood and at the shop rushing around frantically putting things in order, I realise that soon we will be sheltering from ferocious winds that could destroy people's lives.

On the day of Michelle, the ferocious winds cut off the gas by mid-morning; they build in velocity all day. While the electricity still works, Sofía listens to the radio and keeps us informed of changes in predictions on where the eye of the storm is going to hit. The wind and rain are terrifying, the sound deafening. Debris is flung through the air crashing against the house. I hear windows shattering in neighbouring buildings. In the middle of the afternoon, the storm's 135mph wind rips large old trees from the ground downing power lines. The kitchen, opening onto a balcony, is not well protected and the wind tears through blowing things over and covering everything in thick layers of grit. We huddle in the front room, where it's more secure. I'm terrified the roof will be ripped off or the shutters will fly open; the winds rattle them relentlessly trying to break into the safety of our shelter. At 9pm the wind calms and eeriness settles outside. It's pitch black, but I can see balconies hanging precariously as they swing from the tall apartment building next door. This had been evacuated earlier in the day, too dangerous for the residents. Fear plays on my mind as I wonder if we should have evacuated too. What if the building next door towering over us tumbles onto us? Most of the buildings in Havana are old and several stories high. Some decrepit buildings (of which there are many) have a police officer stationed inside ready to help evacuate the residents if necessary. Presciently, all residents along the seaside had been evacuated as the ocean swelled and flooded into the city.

The day after the hurricane I am awakened to hear yelling and the sound of a nearby building crashing to the ground and the background sound of dogs yelping. It sounds like the beginning of a day that will hold a lot of heartache for Cuba. Sofía and I venture out into the street to survey the damage to our neighbourhood. Our street is strewn with large fallen trees. There is much damage and debris. There are electrical wires flapping in the wind. Neighbours talk of the damage in Central Havana, the poorest area of the city. They tell me tenements have collapsed killing five people. The services are out – gas, electricity, telephone, and water – it's hot and windy, which means it's still dangerous to be outside.

Sofía improvises an oil lamp: a glass jar containing one inch of kerosene. A toothpaste tube is cut across the bottom and folded open for support; this stands in the liquid. The liquid soaks up through the tube and we light the wick at the top. She also builds a stove using a tin and lights the fire within. On this makeshift stove she cooks us spaghetti with tomato, capsicum and garlic.

When the electricity comes back five days later people cheer and yell with delight.

Within days of the hurricane, spraying for dengue fever commences. Teams of men move from house to house in Central Havana where stagnant water will undoubtedly produce the next disaster – disease.

In the days leading to the hurricane, I was not sure what to expect. As it was, President Fidel Castro and his government had the planning and infrastructure in place to deal with this disaster. He kept Cubans informed via live television and radio broadcasts. Over 700,000 people were evacuated and although the hurricane damaged 90,000 homes only five people died. Two hundred and seventy thousand people were housed in public shelters receiving organised food and medical attention. I witnessed how civil defence is embedded in the community to deal with a crisis situation.

Responses to natural disasters are always heavily politicised wherever they take place; in developing or developed nations. Commendations from bodies like the United Nations, World Health Organisation and World Bank cite Cuba as a model not only for disaster preparedness, but for healthcare, sustainable agriculture and education. It is tempting at this time to think of this country as some sort of global inspiration. Many people travel to Cuba to see for themselves, how the socialist island ninety miles south of America has continued to enhance its model of social development. This book is about this form of explorative travel.

While Cuba is often depicted as poor and trapped in the 1950s, this is characterised positively in tourist material as part of the yesteryear charm of Cuba, an alternative to the destructive processes of modernity. Certainly, the Cuban government presents its country and society as holding particular values which are an alternative to the individualistic nature of capitalism: people-centred and collective, offering a peculiarly Cuban style of development. Cuba has come to be defined by its socialism and its brand of social development. As this book shows, the people who travel on NGO study tours to Cuba with organisations such as Global Exchange's Reality Tours and Oxfam's Community Aid Abroad Tours, do so precisely because they wish to learn about development within the Cuban socialist framework. This style of encounter is a complex blend of tourist expectation and development aspiration at many levels. It is a growing form of social encounter and warrants careful examination.

Introduction
Cuba: Rhythm, Resilience and Revolution

This book is positioned within the interstices of development anthropology, the anthropology of development and the anthropology of tourism. Located within a space where development and tourism come together in Cuba, I examine the nature of development (anthropology of development) while investigating tourism motivations and experiences (anthropology of tourism). To do both, I worked for a non-government organisation (development anthropology) during the course of this research in Cuba. The boundary between development anthropology and the anthropology of development is blurred necessarily (Edelman and Haugerud 2005), in what follows I draw on my practical development experience together with a post-structuralist critique and contemporary social theory. My endeavour signals a new engagement between anthropology, development and tourism.

Analytically, there is a clear distinction between development anthropology which has expanded since the 1970s and the emergence of a critique of the fundamental premises of Western funded development. It is notable that a growing number of anthropologists are involved with development projects. Development anthropology has been defined as the work of practitioners who design and implement development policies and projects while the anthropology of development is the ongoing analysis of the apparatus of development discourse and practice (Edelman and Haugerud 2005; Gardner and Lewis 1996). Escobar (1997: 505) sums up the differences as follows:

> While development anthropologists focus on the project cycle, the use of knowledge to tailor projects to beneficiaries' cultures and situation, and the possibility of contributing to the needs of the poor, the anthropologists of development centre their analysis on the institutional apparatus, the links to power established by expert knowledge, the ethnographic analysis and critique of modernist constructs, and the possibility of contributing to the political projects of the subaltern.

My book moves between these two domains – working for a non-government organisation (NGO) in Cuba allowed me to utilise historical and critical anthropological perspectives on development (and indeed tourism). Further it allowed me to consider ethnographically the development encounter of tourists in Cuba. This approach has not only been essential to understanding the nature of tourists' experiences and the types of personal development and transformation they bring about, but is also vital in explaining how tourism itself is converging with development. This intersection has implications beyond the tour itself for both local people and tourists. Study tours to Cuba show us new forms of rights-based development in the making as tourists become entwined with, and compelled by, Cuban solidarity. This book highlights how local agency and global forces merge. It shows how Cuban people and tourists potentially emerge from this engagement as influential social actors. This is not to say that all forms of tourism produce this outcome or indeed that all people participating in NGO study tours have these experiences. I will demonstrate however, how most people taking part in the tours had positive educational engagements with local Cuban people and organisations, and how this created outcomes well beyond the specifics of the tour itself.

This book is also positioned within the anthropology of tourism. As noted by anthropologists over the last three decades, tourism is an appropriate topic of anthropological inquiry. Clearly, travel creates contact between cultures and subcultures (Graburn 1980; Nash 1996). Anthropologists have responded to the growing importance of tourism by exploring the social, cultural and economic impacts of this on local people. Anthropological studies have focused on tourists from different perspectives. They include that of personal transition, where the tourist experience is analysed with the help of Turner's existential concepts of communitas and liminas. This book focuses beyond the impacts of tourism to explore tourist experiences within the realms of development. Western ownership and control of tourism activities in developing countries have been well documented and will not be discussed in detail in this book. Rather I approach the notion of tourism as a development paradigm by exploring experiences within NGO study tours in Cuba in order to shed light on the convergence of development and tourism.

Tourism is highly stratified. Individuals do not have equal access to leisure when work is over. Access to different types of leisure is mediated by differences in class, gender, age, wealth and status (Hardy 1990: 553; Rojek 1993: 20-21). Tourism is a source of cultural capital with specific patterns of consumption marking individual identity and belonging to specific social groups (Thurot and Thurot 1983; Munt 1994a). Here the work of Bourdieu (1984) has been most influential. He argues that people's choice and appropriation of symbolic goods and lifestyles are the processes by which they accumulate cultural capital and locate themselves in the social hierarchy. This aspect of consumption explains why specific touristic modes are attached to different social groups at different historical moments (Graburn 1983). Whereas previously (before pre-commercial jet travel) simply being able to travel was a status marker, now the status of travel depends on its style; the destination, travel mode, accommodation type, and activities undertaken. Changes in tourist styles are not random but connected to issues of class competition, prestige hierarchies, generational struggles, and social mobility (Graburn 1983). In exploring such issues we gain a broader understanding of the problems of modernity and identity.

Tourism has been criticised for being based on brief encounters between locals and foreigners that are "artificial, asymmetrical, and unidirectional" (Nash 1981:

468). The organisation of geopolitical inequalities is highlighted by the mobility of Western travellers versus the supposed place-boundedness of the majority of Cuban people. NGO study tours which orchestrate specific forms of social interaction are an attempt to transcend this limitation through the exchange of ideas and knowledge. It is precisely the exchange of development ideas and knowledge that allows this form of tourism in Cuba to be multidirectional. This is because it facilitates the establishment of new networks and international links that in turn assists Cuban solidarity. Moreover, as I explain in the course of this book, they arguably foster a nascent but discernable form of rights-based development.

"If we are to study the nature of solidarity and identity in modern society, we cannot neglect tourism, which is one of the major forces shaping modern societies and bringing (and changing) meaning in the lives of the people of today's world" (Graburn 1980: 64). This is precisely because as Bauman (1997: 93) notes tourists are a "metaphor of contemporary life". Studying tourism is studying modernity because "acting as a tourist is one of the defining characteristics of being 'modern'" (Urry 1997: 2-3) and, importantly, "because tourism is directly responsible for physically exporting the patterns of development associated with modernity worldwide" (Smith and Duffy 2003: 1). Late modern society is internally divided by sometimes contradictory ethical norms. Thus according to Smith and Duffy (2003) there can be no simple answer to whether modernity, development, or tourism are 'good' things. This theme sets the stage for asking precisely how we frame our understandings of the ways tourism and global and social changes are related.

The analysis of NGO tourist experiences is important because they tell us about the ongoing logic of two things. That is, their experiences are shaped by the convergence between development and tourism. Very little has been written about this. The few outstanding recent publications on tourism and NGOs focus more broadly on issues of ethics (Smith and Duffy 2003), the moralisation of tourism (Butcher 2003) and the sustainability of tourism (Mowforth and Munt 2003). Typically studies of development look at the impacts of development policies and programs on local people, but that misses a crucial part of the picture. Once development begins to utilise tourism, the very uncontrollable nature of tourism becomes more apparent. Hence, it is important to base an understanding on what happens in tourist experiences in order to apprehend possible outcomes for development.

NGO study tours highlight an emerging trend for people to incorporate a personal development focus into their travel. These can be understood through tourism paradigms that developed in the late twentieth century, which saw many people go in search of travel experiences that were an alternative to mass tourism (Cohen 1987). The 'new tourism' paradigm established by Poon (1989, 1993) is also germane to a study of NGO study tours as it informs us about important shifts in modernity, the rise of the new middle classes, and how these are reflected in tourism. Finally Butcher's (2003) 'moralisation of tourism' critique presents a cogent framework by which to apprehend the tourist experience in development-oriented tours.

Currently, there is an increasing interest in taking tours operated by NGOs. Perhaps a more ardent form of this is the desire of tourists to travel beyond 'superficial' interactions to more 'authentic' contact with culture by volunteering on conservation and humanitarian aid projects. The 'Travelling with a Purpose' symposium held in San Francisco in 1999 found that many travellers are seeking a chance to be immersed in a community or assist with projects when they travel rather than simply pass through. The symposium discussed an emerging trend of the giving of one's time and talents when one travels; a trend that has many manifestations. These include, activists monitoring sweatshops, observation of local grassroots organisations' initiatives, and assisting with community or research projects. NGO study tours are increasingly both creating and responding to this demand.

Recognising the sheer force of tourism as a global phenomenon of the last several decades has seen development agencies begin to push for reform in the tourism industry. As observers of negative effects of commercial tourism, some development agencies and human rights organisations propose an alternative form of tourism that aims to sensitise tourists to local conditions. "If tourism is to be salvaged, a thorough rethinking and restructuring of the whole industry is called for, taking as its basic premise not profit-making and crass materialism, but the fundamental spiritual and human development of peoples everywhere" (O'Grady 1982: 75-76). Holidays can be an opportunity for people to learn about different societies and to create ties with others; they can be both fun and educational (Benjamin and Freedman 1992).

Perhaps the most recent variant of alternative tourism is where people are using their leisure time for personal growth and educational purposes. In this way, a core aspect of their travel is the incorporation of a learning objective. This can be even further broken down to specific niches. As mentioned, one variety is where tourists wishing to go beyond superficial interactions for closer contact with a culture spend several months or more volunteering on conservation and humanitarian aid projects. This experience affords tourists the unique opportunity to interact and engage over a long period of time with the people and culture being visited.

Another niche within alternative tourism is where people who want a meaningful and educational experience over several weeks (rather than months) look towards NGOs to provide this. Hence, there is an increasing phenomenon where NGOs are more and more involved in tourism. Arguably, this is because organisations at the forefront of development in Third World countries are extremely aware of the issues that Third World cultures deal with on a day-to-day basis and see that tourism is continually growing. Thus, in an effort to minimise the associated negative impacts of much tourism development in so many Third World countries, NGOs are increasingly involved in tourism in various capacities.[1] For example, there are development NGOs that assist tourism operators to create strategies such as responsible tourism guidelines using in-country operators and products rather than multinational corporations' hotels and restaurants. Some development organisations help to establish links between in-country operators, overseas tour operators, government bodies and non-government organisations in an effort to create a cohesive network working towards informed and responsible tourism practices and policies. Other organisations actually have subsidiary arms that develop tours focusing specifically on the development issues of a particular country and provide tour participants with opportunities to meet with local in-country grassroots organisations, meet local people to discuss these issues with and visit local development projects. It is this last grouping that I am concerned with.

As explored throughout the book, the involvement of NGOs in tourism appeals to some people's ethical perceptions of 'responsible', 'sustainable', or 'educational' tourism. Butcher (2003) refers to this ethical concern as the "moralisation of tourism" where tourists engaging in so-called responsible forms of tourism consider themselves as morally superior to other tourists. Butcher warns us that such new forms of tourism are not necessarily benevolent and should in fact be subject to the same criticism as problems associated with mass tourism. Throughout the course of this book I address Butcher's critiques of moral tourism and extend the concept in a way that sheds new light on specific tourist experiences.

Book Overview

This book builds on research by Mowforth and Munt (2003) and Butcher (2003) into the tourism-development nexus whereby development organisations are increasingly becoming involved in tourism. The intent of this book is to inform critical attention to convergences between development and tourism. Tourism operating within the framework of development agencies brings together tourists and host communities where tours are designed to impart firsthand knowledge of the political, social, economic and environmental conditions that impact on local development. Thus, an emerging area of 'socially responsible' and 'educational' tourism is observing local grassroots organisations and community projects in the countries visited. For example, Oxfam Community Aid Abroad Tours (*OCAAT*), a subsidiary of Oxfam Community Aid Abroad, offer small group study tours, which visit the community projects supported by Oxfam and other grassroots organisations. Such tours typically include in-country presentations about the role of community projects and the opportunity to visit and observe some of these.

Tourism organised under the auspices of development agencies is considered by many people as a responsible form of travel because the study tours aim to

---

1   Fred Hollows Foundation, Oxfam, Charities Aid Foundation, Childwise (Ecpat), Detours Abroad.

educate tourists through visiting aid projects, attending presentations about healthcare, education, environment, gender, sustainable agriculture and other development issues pertinent to the country. This book highlights to what extent development-focused tourism acts as an instrument of social and cultural understanding by providing facilities for the acquisition and exchange of information about the social, cultural, political and economic endowments of local people. The tour operators argue that tourists will return to their communities and disseminate this new knowledge and that increased understanding will result in changed attitudes and behaviour. It is hoped this will lead in turn to a more just and equitable relationship between nations and achieving a mutually beneficial relationship between development and tourism. These are big claims. I question how meaningful such experiences are and what ongoing implications they have for development in the country visited.

Thus the core aim of this book is to examine the tourist experience in terms of how the development issues of the host culture are interpreted and understood within development-oriented tours. I present an ethnographic account of NGO study tours and participating tourists; their motivations, experiences and outcomes. In this way we can begin to understand how tourist interactions with community projects affect the lives of tour participants in relation to forms of emotional or intellectual transformation. The empirical aspect of this research also focuses on the ways in which cultures are experienced by tourists in development-oriented study tours. The chapters collectively argue for an understanding of the extent to which such tours contribute to the enhancement of cultural awareness and, ultimately, some form of commitment to solidarity and development causes.

I'm mostly concerned with the moral tourism-development nexus that involves Western tourists actively using their holidays to learn about development issues. It is through this process that tourists achieve a sense of agency. As I elucidate the increase of a developing tourism niche, which focuses on an exchange of knowledge, undoubtedly has positive implications for a host country, leading as it does to the creation of global networks and solidarity.

The tours have the power to affect participants in positive ways that encourage people to be more active in a growing social movement of foreigners supporting Cuban solidarity. What we will see produced here is rights-based tourism. I will establish that notions of touristic transformation are complex forms of identification from the rejection of mass tourism to refutation of neo-liberal globalisation. Tourists demonstrate identification with certain interests. Transformation is qualified by the tourists' affirmation of values espoused in the West – such as social equity and collective community – but no longer considered valid with the collapse of the Soviet Empire. Cuba forms an example of a model that tourists are compelled to study, support, and promote.

Solidarity through tourism can therefore be considered an important tool for development agencies, social movements and non-government organisations. Specifically in terms of new and explicit ways of promulgating issues of rights, social justice and good governance. In this way, solidarity connects directly with rights-based development. We can effectively envisage solidarity as a means for tourists to participate and act as agents of change in the development process. Indeed it is a novel means through which Cuba has developed a way to partially overcome the economic and social constraints of the US blockade. It acts as a new form of global coalition and interconnectedness that builds on previous alliances that have since dissolved. This was the case with the Soviet Union where Cuba engaged in cultural exchanges with nations who were politically sympathetic.

Oxfam and Global Exchange

Two NGOs that have been heavily engaged in tourism are the Australian Oxfam Community Aid Abroad and the San Francisco based Global Exchange. This book is based on my association with the former Oxfam Community Aid Abroad Tours (*OCAAT*) and Global Exchange's Reality Tours (*GERT*) and their tours to Cuba. These two tour operators are subsidiaries of development and human rights organisations that are devoted to 'responsible' tourism in developing countries. A central focus of their operations is to educate tourists about the development issues that people are faced with and what local and international efforts are doing to alleviate and improve them. The idea is that by educating tourists and providing them with opportunities to meet and discuss these issues with local people they will a) minimise the negative impacts typically associated with tourism; b) disseminate what they have learnt; and c) work towards supporting efforts to help these countries.

Oxfam's *OCAAT* and Global Exchange's *Reality Tours* (*GERT*) were considered appropriate organisations for this research because of their commitment to development issues and 'responsible' travel. *OCAAT* is a subsidiary body of Oxfam Community Aid Abroad in Australia and *Reality Tours* (*GERT*) is a subsidiary of Global Exchange. Both organisations aim to educate the travelling public about development issues through 'sustainable', 'responsible' and 'educational' tours. The guiding principles of these forms of travel are to understand the culture visited, to respect and learn from the people who are hosting the visit, and to tread lightly on the environment. Tours by both organisations are hosted by the Cuban Institute for Friendship of the People (ICAP) and include meetings with, and visits to community projects with a combination of non-government grassroots organisations, government agencies, academics, doctors, artists, and the like. This is combined with a social itinerary to provide an understanding of contemporary Cuban life.

For example, four *OCAAT* tours that I co-ordinated focused on a broad selection of topics with each day devoted to one or two issues. These included visits to community projects and meetings with many people. Tour participants visited a Cuban women's project and learned about women's lives in Cuba today; they met staff and students at an art and music school to see how Cuba teaches its young artists and musicians; visited a school and discovered how world class literacy rates of over 90 per cent have been achieved; visited a provincial health clinic exploring how the government has established such a healthy long-lived population and visited a permaculture project to see sustainable agriculture. The idea behind these is that Cuba has a national program of social development that can be identified through these (and other) institutions. Hence art, schools, and health clinics, all tell us the story of Cuba. The three study tours I participated in with *GERT* have specific foci including a Women's Delegation, Sustainable Agriculture and Cuba at the Crossroads. The Women's Delegation was hosted by the Federation of Cuban Women (FMC). Tour participants had opportunities to learn about the contemporary effects of tourism on women and specifically the reappearance of prostitution. Among other things we were told about the Family Code and Cuban society's attempt to integrate women fully into the workforce and to equalise housework and child care between men and women while guaranteeing equal pay. The study tour visited *La Guinera*, a community development project inspired and led by women. The Sustainable Agriculture tour highlighted Cuba's current engagement in the most comprehensive conversion from chemical to organic agriculture any nation has yet attempted. The study tour met with farmers, professors and other experts in the agricultural industry, visiting sugar cane, rice, vegetable, root crop, animal production sites, and green medicine gardens. The tour also visited community run urban gardens, farmers' markets and agricultural co-ops in various provinces. As a socialist country, they face a host of challenges and the Cuba at the Crossroads tour was designed in an attempt to give an overview of the various aspects of society: economy, government, health care, education and so on.

Reflections on Cuba

> Reflexivity requires an awareness of the researcher's contribution to the construction of meanings throughout the research process, and an acknowledgment of the impossibility of remaining 'outside of' one's subject matter while conducting research. Reflexivity then, urges us to explore the ways in which a researcher's involvement with a particular study influences, acts upon and informs such research. (Nightingale and Cromby 1999: 228)

My time living in Cuba as an anthropologist in the early 2000s was at a moment when the country was emerging from a decade-long depression. I conducted fieldwork in Cuba between October 2001 and March 2003 and returned there in early 2004 for another several months. I co-ordinated tours for *OCAAT* and participated in several *GERT* tours. At times, it was extremely challenging to manage and negotiate the demands between my roles as Oxfam Co-ordinator, Global Exchange tour participant and ethnographer. My Oxfam role required me to research and plan for upcoming tour programs. I worked on research and development of the program liaising with Cuban grassroots, development, and government organisations throughout the duration of my time in Cuba. This position gave me a privileged opportunity to explore the realm of this tourism development nexus.

My role on the study tours changed according to which organisation was operating the tour. For example, with the *GERT* study tours I was a participant and researcher. This saw me introduced to all other tour participants as a researcher interested in the study tours. With the *OCAAT* study tours I was tour co-ordinator, participant and researcher. The Cuban tour leaders would introduce me as the co-ordinator and I would then explain that I was also conducting research. Hence my roles varied from that of a passive participant observer on the *GERT* study tours to that of actively participating in *OCAAT*'s. This was because my role with *OCAAT* involved designing and co-ordinating the itinerary of presentations and project visits for the study tours. Thereby, significantly influencing the types of experiences to which people would be exposed.

My most demanding role while conducting my fieldwork in Cuba was as the Oxfam Tour Co-ordinator. I was not just a disinterested researcher, but a professional with experience and investment in the kind of tourism that I was researching. I was responsible for the research, design, implementation, and sourcing of new contacts within development and grassroots organisations and overall co-ordination of the Oxfam Community Aid Abroad Tours in Cuba, and ensuring that the interests of tour participants and *OCAAT* were met. It was my role to oversee commitment of the study tours to the principles of responsible travel as espoused by *OCAAT*. My position included acting as a resource person, assisting with orientation and briefing of tour members, translating language and cultural differences, and, providing advice and recommendations to the Director on improved practices and initiatives that would enhance the development of future tours. The role required me to be knowledgeable about Oxfam Community Aid Abroad, grassroots community development, local in-country knowledge, to speak the local language, be a problem solver, demonstrate a commitment to social justice, to show sound judgement and to have the ability to deal with unforeseen circumstances.

Living in Cuba, my identity constantly switched between that of tourist and temporary resident. I had secured a temporary resident visa, which gave me access to local prices in pesos, but when I was 'on tour' I paid tourist prices as part of a tour group. Arriving in Havana, I quickly realised that Cuba is profoundly *machista*. Local *Habaneros*, young and old, tirelessly cast *piropo* (witty compliments) as I walked past. This was despite my usually careless appearance compared to the groomed beauty of Cuban women who take great pleasure in hearing these compliments. As an Australian woman I was used to walking down Australian streets in complete anonymity without cause for self consciousness as men in Australia do not typically make such comments. However, as the locals became familiar with me, and as my Spanish improved, I no longer felt so conspicuously foreign. In the Cuban way, I was given a name by the locals that reflected a prominent physical characteristic. This was typically the colour of your skin, your weight, or ethnic background such as *Negrita* (black girl), *Gordita* (fat girl), or *China* (Chinese girl). In my case it was *Blanquita* (white girl) because of the paleness of my skin. Walking down my street the neighbours would often call out "*¿oye Blanquita cariña que vola?*", loosely translating to "How are you darling"?

The Spanish colonial house I moved into in the neighbourhood of Vedado was nestled between crumbling towering apartment buildings near the *Universidad de la Habana*. Its pretty façade of pink and brown belies the fading grandeur of a home filled with light, a grand chandelier, antique furniture and a magnificent collection of framed butterflies in the *sala* (front room), unreachable ceilings, light blue wooden shutters and decorative wrought iron. The centre of the house, typical of Moorish architecture, opens to the sky with a narrow balcony that connects the kitchen, bedrooms and bathroom to the *sala*. Reddish brown wrought iron stairs lead, simultaneously, down to where the family resides and up to the roof terrace. The house had a lovely light airy feel. The bright blue shutters that we nailed shut during Hurricane Michelle's onslaught were more typically opened against white walls and antique Spanish tiled floors. Potted plants along the balcony and pretty orchids perched on the baluster attracted little black birds with yellow streaks. Sofía's house, where I spent a great deal of my time in Cuba, had the most wonderful atmosphere. It was always alive with *Abuelita* (Grandmother) constantly chattering away, Frida keeping every inch of the house clean and Sofía busy attending to all kinds of pressing matters concerning the travellers, students, friends and family in and out of the house all day long. Not to mention the peaceful presence of *Abuelo* (Grandfather) reading the daily paper and taking his siesta on the antique settee in *la sala* or down in the quiet dark sanctuary of his bedroom. For me, Sofía's house and the street represented a microcosm of the lively chaotic nature of Havana. The daily problems of many local people flowed in and out of that house; a house in need of attention and care just like the people, but an enduring resilience of both exists. Outside in the street, the old *Habaneros* sat on the front doorsteps chatting and watching the goings on at the local market across the road as Mario, the mechanic next door, worked on an endless stream of old American cars outside his house under a large tree whose thick branches he used to make pulley systems to lift heavy engines from the car he was working on.

Sofía's house sits in a once wealthy neighbourhood. It is a beautiful example of Moorish architecture. Its current use is also an example of many contemporary Cuban households – several generations living under the one roof, sharing a bedroom and beds in one section of the house while the main house is rented to foreigners for hard currency. In this house, Sofía, her parents, her niece, and a housekeeper lived downstairs in the basement where they shared a tiny bathroom, kitchen and bedroom. Meanwhile the three bedrooms of the main house were rented to foreigner students, tourists and researchers such as me, who also had use of the main bathroom, kitchen and *sala*. While this was legal, each rented room incurred a fixed tax; many Cubans furtively rented more rooms than they claimed to the authorities. While living there, on more than one occasion the authorities would do random inspections to enforce tight controls over these new Cuban entrepreneurs. We would run around frantically, but discreetly, while Sofía delayed the authorities as we rearranged two of the bedrooms to look like they were lived in by Sofía and her parents while the third bedroom was rented to a foreigner. This entailed the quick transfer of all foreign and expensive looking products and clothing into one bedroom and then hiding behind the shower curtain while the authorities did their inspection of the bedrooms. The fines for Sofía if caught were astronomical by Cuban standards and she would have lost her licence to rent a room. But at the time that I was living in Cuba it was a risk that many Cubans took because the rationed food and government income was not enough to get by without supplementing from the black market, and that required hard currency.

Living with a local family was crucial for improving both my Spanish and my understanding of such social, cultural and political issues, and in accessing areas of life that would otherwise have been closed to me as a short stay tourist. Sofía generously introduced me to a range of people and facilitated many of my early research inquiries. As she was operating a *casa particular* (private house with rooms for rent) I learned how many Cuban people use the tourism industry to earn their livelihoods, precariously teetering between the burdensome bureaucracy of socialism and the tantalising profits of capitalism. Living with a Cuban family also gave me insight into how many people use the black market to supplement their income. Every week ladies would turn up at our front door with a small backpack. They came in to the *sala* and took a seat in one of four rocking chairs, ubiquitous throughout Cuba. They sold homemade cheese, coconut sweet, ham, coffee and the like. They were equipped with an old fashioned scale with a hook to weigh the pounds you needed. One lady from the neighbourhood came several times a week to rent her pirated videos to us on the *mercado negro* (black market). Many were American movies with Spanish subtitles. On my second visit to Cuba, this lady had been denounced by someone in her street and she was subjected to an inspection. The authorities took her video player, television, and many videos. Apart from losing a large part of her business she would have paid a *multa* (fine) of 1500 pesos. At that time she had around 200 videos out on rent so she was still able to operate her clandestine business. Staying with Sofía and her family and getting to know the local neighbours is not something a tourist would have had the luxury of experiencing in the early years of the revolution. It is indicative of a changing tourism scene in Cuba.

Figure 1 Surveying the damage caused by Hurricane Michelle in our street

Figure 2 The house where I lived in La Habana

Figure 3 A market place

"The Evil We Have to Have" – *El Turismo en Cuba*

Fidel Castro has famously referred to tourism as "the evil we have to have". This statement highlights his awareness of the unpredictable nature of tourism and signals his intent to make use of it on his own terms. As I witnessed, since 2001, and will detail throughout the book, policies for tourism in Cuba have brought ongoing changes. Changing styles of tourism tend to reflect the changing political and economic situation in Cuba. Throughout the twentieth century, Cuba as an international tourist destination has undergone recurrent transformations. Prior to the 1959 socialist revolution Cuba was the leading tourist destination in the Caribbean. Following a slump in tourism during the revolution, international tourists began returning in 1975. This initially comprised large numbers of Canadians. Since the 1990s tourism has been placed at the forefront of development policies and arduously pursued. Currently, Cuba's tourist industry is one of the fastest growing in the world, providing the country's largest single source of foreign exchange. Varadero on the north coast is one of the largest tourist resorts in the Caribbean. Most tourists today come from Canada, England, Italy, Spain, Germany, France, and Mexico.[2] In discussing the turbulent development of tourism in Cuba, Bleasdale and Tapsell (1994) usefully identify three distinct phases: pre-revolutionary tourism (up to 1959), revolutionary tourism (1959-1987),

---

2 Oficina Nacional de Estadisticas 2008.

14

Development Tourism

and tourism since 1987. I have identified a further phase, namely contemporary development tourism.

Varadero was the beach resort for middle-class Cuban families. However, after the War of Independence with America (1895-98), Cuba became the playground for wealthy Americans who purchased cheap beachfront property thus marking the beginning of the international tourist trade (Jimenez 1990). During the Batista regime (prior to Castro's revolution) Cuba was "the tourist hub of the Caribbean, particularly during the 1930s and just after the Second World War when many thousands of visitors came to the island" (Hinch 1990: 215). The development of tourism was concentrated in the capital city Havana and Varadero on the north coast. Until 1959 Cuba was the main Caribbean destination for tourists from the US. In 2002, Miguel Figueras, then advisor to the Minister of Tourism, stated that tourism prior to the 1959 revolution developed in close relation to the rise of Las Vegas in America, as there were strong Mafia and gambling connections between Las Vegas and Cuba. Tourists were predominantly Americans interested in gambling and prostitution at the Mafia owned casinos.

Che Guevara and Fidel Castro's revolution associated tourism with the exclusive and indulgent nature of the upper echelons of society during the Batista regime. It was this capitalist enclave that Castro intended to eradicate with the move to socialism. With the move towards communism, the US outlawed Cuba as a travel destination for its citizens thereby eliminating over 80 per cent of the foreign tourist market. A new form of tourism was ushered in. Domestic tourism became the focus, with an emphasis on socialist ideals characterising tourist activities throughout the 1960s. For example, tourism developed strong educational and political aims and Cuba made economic alliances based on tourist traffic with politically sympathetic nations. This established a new international, albeit limited, tourist trade with the former USSR, Czechoslovakia and East Germany under the banner of cultural exchanges (Bleasdale and Tapsell 1994). Although there were tourists from other countries at this time, the State limited them to group tours organised for the purposes of reinforcing socialism (Hollander 1986). Typically, tourism that advocates a strong political focus has a dual purpose for the host country. In Cuba, it presented an opportunity to demonstrate the benefits of socialism, an objective the government felt it needed to achieve on an international level due to its political struggle with the US. It also meant that the government could control the activities of visiting foreigners by imposing tight controls over the itineraries of tourists, thus minimising the risk of ideological contamination of the Cuban population. There emerged a separation between foreigners and the local population as opportunities for international tourists to meet locals were reduced by the government forbidding local people to use tourist facilities. From the mid 1970s Canada was targeted by the marketing of winter package tours and tourism was further expanded by a relaxation of the ban on American citizens travelling to Cuba during the Carter presidency (Bleasdale and Tapsell 1994: 101).

In the late 1980s, the government positioned tourism as the leading replacement industry for its economic survival. However a lack of capital hindered the process, and in addition many of the resources needed to develop tourist facilities had to be imported at great cost. The introduction of new legislation in the form of a changed constitution and the implementation of other mechanisms were aimed at generating external finance to overcome these problems. In the recent decades the government amended legislation to permit joint ventures with overseas corporations in the construction and refurbishment of hotels and tourist attractions such as *Habana Vieja* (Old Havana) precinct. Correspondingly, Cubanacan, CIMEX and Havanatur, State owned tourism enterprises have been set up to facilitate tourist based joint ventures. In addition, the government established a host of organisations to ensure the provision of improved services for tourists. Another measure taken by government in bringing Cuba up to international tourism standards is the establishment of hotel training schools. It is evident in Cuba today that the tourism industry is increasingly providing what are considered to be very prestigious jobs for highly educated and multilingual people (Bleasdale and Tapsell 1994). It is not uncommon to find lawyers and other professionals who speak fluent English driving tourist taxis. They can earn more *divisa* (hard currency) through working in an associated tourist industry than through their trained professions. And it is not uncommon to find people like Sofía renting rooms in their houses while simultaneously working in other professions; Sofía is an editor at Radio Habana.

The collapse of the Soviet Union, Cuba's sugar daddy, ended the economic subsidies and barter arrangements. This caused the chronic shortage of oil, suspended most sugar exports, and led to food shortages. Castro announced that Cuba was in a 'Special Period' in a time of peace. Thus Cuba's growth strategy was an effort to overcome the various difficulties it had faced over the last several decades and to rapidly expand the three sectors believed to hold the most potential for an injection of hard currency (Cross 1992): the food program through sustainable agriculture; biotechnology and pharmaceuticals; and tourism, the most critical (but also a potentially threatening) industry for the survival of the socialist State (Bleasdale and Tapsell 1994: 103).

It seems ironic that the government has encouraged the growth of tourism, because inevitable ideological tensions created through the implementation of a capitalist industry could potentially undermine socialist structures. Without a doubt the government is faced with the dilemma of facilitating co-existence between Cubans and international tourists in ways that minimise potential negative social, cultural and ideological impacts. But as Stephen Wilkinson (2008: 980) points out:

> Indeed, the real irony might be that tourism in Cuba has been its saving because it has enabled the socialist system to survive. It is this system, with its macroeconomic control of the economy by a strong interventionist State that mitigates effects that tourism might otherwise have. If tourism development had been left to uncontrolled market forces, then the problems that have occurred would have been a great deal worse.

Certainly contemporary tourism developments display some of the characteristics that typically lead to negative social impacts – geographical concentration, resort enclaves, separation of the spending of local and foreign populations, and dependency on a global market. According to Bleasdale and Tapsell (1994), tourism in developing countries has been typically characterised by high levels of dependency on the external markets, foreign capital and imported human resources of multinational corporations. In Cuba, tourism development displays some dependency characteristics through the increase of external financing of joint ventures. Similarly, tourism development in many Third World countries is largely enclavic and where tourists are intentionally separated from the local people. Enclavic tourism generally stems from multinational corporations' unilateral financing of beach resorts to accommodate package tourists. International rather than local currency is spent within the confines of these resorts further reinforcing the separation. The enclavic tendency remains evident in Cuba as the government introduced the US dollar in 1993, withdrew it in 2003 and, has since 2006, replaced it with the Cuban Convertible Peso. In addition, tourism development is concentrated within designated regions in an attempt to limit the sociopolitical impact of numerous visitors. This has inevitably led to the development of tourist enclaves such as Cayo Largo on the north coast, increasing the geographical concentration of tourism and greatly limiting the benefits local people gain from tourism spending. However, tourism also thrives in Havana where tourists engage with locals in bars, in the street, along *el Malecón*, and in cafes, signifying the uncontrollable nature of tourism. The NGO study tours break down the enclavic nature and build on the socialist support philosophy.

The Cuban government has implemented a 'niche' tourism marketing strategy to benefit the country's economy. Niche types of tourism that the industry currently spans include cultural and heritage, socialism, beach resort, conference and business, ecotourism, special interest, and one not supported by the government – sex tourism (Tribe 1999). Health tourism is growing in Cuba with an increasing number of visitors taking advantage of the country's advanced and relatively inexpensive medical technologies, which are a notable source of hard currency for the Cuban government. In 2008, 10,485 tourists visited Cuba for health tourism.[3] These visitors arrived in Cuba as health tourists seeking help for a wide range of medical conditions and ailments for which Cuban medical practitioners have developed treatments. *Servimed*, the network of specialist medical and health centres, offers both medical services for clients and training at all levels for health professionals. Clients enjoy access to the most advanced medical treatments, modern technical equipment and unique standards of comfort as well as enjoying the beauty of the countryside. Centres offer treatments for conditions such as: hypertension, pigmentary retinosis (or night blindness), Parkinson's disease, psoriasis, deformities of the spinal column, bone tumours, paralysis, and

---

3   Oficina Nacional de Estadisticas 2008.

rheumatic diseases. Health tourists also take advantage of *Servimed's* therapeutic communities for the treatment of drug addiction and stress.

Ecotourism represents another growing niche in the tourism sector that the Cuban government continues to invest in. This includes environmentally friendly facilities, which offer, for example, solar water heating, the use of biodegradable detergents and biological pesticides, water recycling or nonchemical waste treatment facilities. A number of Cuban properties – the Sol Palmeras in Varadero and the TRYP in Cayo Coco, are two examples which have been promoting their environmentally friendly facilities and practices for the past decade. Cuba offers many opportunities for ecotourism that are already lost to other parts of the Caribbean. This is due partly to military security zones against potential attacks by the United States along much of the north coast including the northern cayes. Also many of Cuba's best tourist attractions remained nearly pristine because of the island's small tourism industry through most of the post-Revolution period. Cuba also possesses a fairly well established legal, administrative and physical foundation for ecotourism development compared with many developing countries. The National Commission for Protection of the Environment and the Rational Use of National Resources was created in 1977. In 1994 the Commission's functions were incorporated in to the Ministry of Science, Technology and the Environment. In 1981, the *Campismo Popular* (People's Camping) agency was created. This agency developed over 100 designated camping areas across the island including many in tourism 'hot spots' such as Caye Coco and Caye Guillermo. Tour agencies such as Rumbos developed specialty tours for these areas with an ecotourism orientation. In addition, four areas in Cuba have been designated as 'biosphere reserve' by UNESCO (United Nations Environmental, Scientific, and Cultural Organisation) to be protected and developed in accord with UNESCO's international principles for such reserves. Cuba has a well established record for environmental protection, which includes an Environmental Impact Law that requires environmental impact assessments be undertaken on all tourism projects before they are approved. In its 2006 Living Planet Report, the World Wildlife Fund recognised Cuba as the only country worldwide that is developing in an ecologically sustainable way.

Increasingly visible on the streets of Havana, sex tourism is one niche the government has attempted to eradicate. At its outset, the Cuban revolution attempted to outlaw prostitution and create improved social conditions for women, which was aimed at their reintegration into employment and full gender equality. Prior to the collapse of the Soviet Union, realizing these goals had been fairly successful to some extent, but since having to reintroduce tourism, the Cuban government claims that prostitution has re-emerged. Certainly the 'dollarization' of the Cuban economy (the legalisation of US dollars) which involved the creation of dollar stores *Tiendas Americanas* for tourists (and these days for Cubans who have access to hard currency), restaurants, nightclubs, tourist taxis and hotels, along with the tightening of the embargo led to a proliferation of prostitution. The need for many Cubans to gain access to the dollar has resulted in commercial sex work becoming increasingly endemic as Cuban girls are drawn to the capital city to earn dollars. Some women are drawn to commercial sex work for the certain lifestyle it affords; they have access to the alluring clubs, bars, restaurants and clothes that are not accessible to local people, but only for foreign tourists. Locals are prohibited from enjoying tourist designated areas such as many clubs, bars, hotels, and restaurants unless they are accompanying tourists or work there.

But perhaps the most distinctive trait of sex tourism in Cuba is that prostitution is not explicitly visible in the public domain because there is no organised sex industry with brothels, sex shops or advertising. Sex tourists are approached in much more subtle and casual ways. This informal and unorganised nature of sex tourism in Cuba makes it difficult for the authorities to eliminate its roots. Consequently one way they aim to target it is to educate young girls about the social, health and legal risks of prostitution through national social service programs with the help of the neighbourhood divisions of the Federation of Cuban Women.

Clearly the informal nature of the sex industry in Cuba appeals to many foreign sex tourists precisely because many are not just looking for sexual encounters from their liaisons in Cuba. They covet companionship from women, a temporary girlfriend with whom they can engage in interesting conversation, have dinner with, take dancing or travel with through Cuba. The Cuban women play the roles of tour guide, girlfriend, companion or dancing partner (Wonders 2001). Typically, Cuban women who work on the edges of commercial sex are well-educated and multi-lingual with many having professional careers, unlike commercial sex workers in other countries. The tourists expect to be "pseudo boyfriends" and they do not want to be treated as clients or customers (Wonders 2001: 563). Given the increasing phenomenon of globalised sex tourism, Castro's attempts to address sex tourism from the inside out is perhaps the only way to grapple with sex tourism in Cuba. The Cuban government cannot control the global forces that shape the production and consumption of sex tourism but it does attempt to educate its female population about the ill effects of prostitution.

The most defining characteristic of tourism in Cuba is its socialist nature. It has since the 1960s involved highly centralised planning and State ownership and as with Eastern Bloc socialist countries, Cuba concentrated more on domestic tourism until the 1970s. Excessive bureaucracy, low standards of service and poor infrastructure (typical of socialist countries) has meant limited interest in Cuba as an international tourism destination. Hall (1984) states that set in a Stalinist framework, the objectives of such tightly controlled packaged tours were to maximise hard currency earnings whilst not sacrificing socialist values. Hall tells us that the political tours of Cuba from 1959 had many features of authoritarian tourism and that these carry over into contemporary tourism as it is still very centralised in terms of planning. Ironically, in many ways the enclavic tourist resorts of the free market forms of tourism lend themselves to retaining some of this control (Bleasdale and Tapsell 1994).

Socialist Cuba's reliance on international tourism has necessitated some free market measures. While the government largely controls the tourism industry, it's initiated changes in legislation allows foreign investors to own tourism enterprises and engage in joint ventures in an effort to bring benefits both to the investors and Cuba (World Trade Organisation 1998). The government states that its decision to allow foreign investment "is an expression of the political will to do whatever is necessary to boost national development without sacrificing national independence and sovereignty".[4] The objective of the joint ventures was to open the door of the tourist industry to an indispensable source of foreign capital, technology and markets. Ideally, this would help to lift Cuba out of its economic hardship while continuing to support the fundamental socialist structures of 'cradle to grave' social security, free education and free healthcare. Foreign investors are subject to Cuban taxation, and so augment government revenues and foreign exchange. The government also exercises an indirect influence over these ventures by controlling the provision and remuneration of the labour force.

Through the increase in international tourism, many Cuban people invent ways to gain quick profits via the free market. Tourism has provided the vehicle for accessing hard currency, whether directly as a tourism employee or through an associated industry. Jobs directly related to tourism are prized positions that are used to reward government supporters. People in professional jobs such as teaching can earn as little as $12US a month and sometimes supplement this income through illicit activities with tourists. While begging on the streets is minimal there is an ever increasing number of *jineteros* (hustlers or literally jockeying on the backs of tourists) who, for example, can earn as much as $100US a week selling contraband cigars to tourists. In Havana local people approach tourists to sell them cigars, tout restaurants or private rooms, or they drive a *maquina* (old American cars used as taxis charging ten pesos, equivalent to approximately 50 cents US). As I mentioned earlier, some locals also rent out their accommodation to tourists (*casa particular*), a practice legalised by the government in 1996, while others transform their lounge rooms into *paladares* (small restaurants) where they are allowed to seat up to twelve people at a time. Although tourism has helped to spread wealth there is a clear disparity. Most skilled and highly trained people are not the ones making the money, but those who can earn the tourist dollar. The government is acutely aware of this and although it has legalised some private enterprise, it has increasingly imposed huge taxes on all forms of private business, forcing many people to close their doors or operate illegally.

Investigating Experiences of an Ephemeral Nature: The Methodological Process

My previous work with NGOs has lead to a long term interest in the relationship between development and tourism; two phenomena that have grown independently yet often exist side by side as well as overlap. The gradual convergence between the two raises many questions concerning the agendas of development agencies,

---

4 Foreign Investment Act pamphlet 1995.

the experiences afforded to travellers in a learning capacity, and the broader consequences for development in Third World countries. Interestingly, both development and tourism have been criticised for impacting negatively on poor people of the Third World. Perhaps the increasing convergence of the two domains can be seen as an attempt to move towards better and more efficient practices for addressing development issues of participation, sustainability and poverty reduction.

Traditionally the assumption that has driven most anthropological research is that it should be about 'other' societies, colonial or ex-colonial (Muetzelfeldt 2002). Recently, there has been a move by anthropologists towards research that is considered by many ex-colonies as being in their national interest, although such interests are often much contested. At the same time there has been a shift toward research at home where anthropologists study aspects of their own society. This shift points to the unexplored limits of what counts as anthropology (Muetzelfeldt 2002). This book reflects both these shifts in anthropological research. It addresses issues of development and tourism in Cuba, both being key concerns of national interest, and at the same time it investigates experiences of more ephemeral nature where Western tourists learn about these issues. While the research was conducted outside my own society, in part, it investigates experiences of people residing in Australia and the somewhat similar societies of Britain and America.

The research undertaken for this book also reflects what Marcus (1995: 95) referred to as multi-sited ethnography:

> An emergent methodological trend in anthropological research that concerns the adaptation of long-standing modes of ethnographic practices to more complex objects of study. Ethnography moves from its conventional single-site location … to multiple sites of observation and participation that cross-cut dichotomies such as the 'local' and the 'global'.

As much earlier exemplified by Malinowski's *Argonauts of the Western Pacific*, to follow mobile people is a compelling example of a multi-sited ethnography. Pilgrimage, migration, and tourism are excellent domains for this style of ethnography, which I utilised while living in Cuba. I tracked the progress of the study tours and the meetings and projects they encountered by travelling with them as they moved from place to place.

Throughout I have adopted a qualitative approach. Henderson (1991) states that qualitative research is defined by an emerging design in a real world setting and is preferred when the outcomes sought include the identification of patterns and the development of theory. Qualitative research seeks a depth and richness in the data gathered that allows for a more comprehensive understanding of a few individuals. Two decades ago Cohen (1988a) claimed that the use of qualitative methods in the area of tourism research was scarce and often lacking in rigor but that the few studies adopting this approach had led to some of the most significant contributions to the field.[5] I adopted a primarily qualitative approach in order to gain an in-depth insight into the experiences of the respondents as they perceived them (Sarantakos 1993). Ethnographically I investigated the experiences of the participants of study tours conducted by the ancillary arms of an international development agency and a human rights organisation. Over a period of several years, the research involved the use of different methodological tools by combining formal recorded interviews, direct participation and observation, literature and document analysis, qualitative email questionnaire implementation and analysis, informal conversations and introspection. My adoption of these methods reflects the particular empirical objective of the research – to understand the experiences of learning about development in Cuba within a tourism context. The ethnographic data collected amongst NGO study tourists is not isolated from other information sources; rather it is linked with data collected from the grassroots and development organisations and government agency authorities that came together in Cuba.

A review of the tourism literature indicates that the nature of tourism (being a process occurring over a number of regions) leads to concerns about the most appropriate times and locations to conduct research involving tourists. It is claimed that research occurring at only one or two points in time leads to gaps in knowledge of the tourism experience (Jafari 1987). It has been argued by Stewart and Hull (1996) that research designs measuring current rather than past mental states are most accurate. They questioned the ability of individuals to recall, recreate or reconstruct past mental states and therefore suggested the use of multiple collection, in-situ sampling methods. Various issues arise however if the researcher intends to collect data during all of the pre-travel, on-site and post-travel phases. Ethical issues and time and financial constraints meant that the data collection methods for this research were designed with no data collected from the prospective tourists during the pre-travel phase. This was largely because the organisations facilitating the research simply did not have the resources to undertake the required ethical procedures in gaining consent from tour participants. Although collecting data during the post-travel phase has been criticised I found it had a number of benefits for this research into tourism. First, a post-tour questionnaire aided in minimising the possibility of impacting on tour participants' experiences. Second, the criticisms made regarding post-travel data collection were carefully considered and then addressed through a combination of participatory observation, interviews and informal conversations during the on-site travel phase, implementation of email questionnaires during the post-travel phase and the collection of supporting evidence during the on-site and post-travel phases. It was hoped that this combination of methods (triangulation) would enhance the reliability and validity of the research. The following section outlines the methods of data collection that were employed during the on-site and post-travel phases.

One potential difficulty that I anticipated and that has been observed by other researchers is that it can be problematic to conduct serious research about topics

---

5   See Boorstin 1964; McCannell 1973, 1976; Turner 1973, 1974 and with Ash 1975.

that are "so bound up with leisure and hedonism" (Crick 1989: 311). Further, since tourists are on holiday, even if it is a study tour, they often do not want to answer too many questions (Oppermann 1996: 92). Changing one of my core research methods to post-tour questionnaires aided in addressing this potential dilemma. Rather than interviewing people in Cuba while they were still immersed in the tour experience, absorbing and processing the information they were receiving each day, I minimised the possibility of impacting on their experiences through participatory observation during the tour and by implementing the questionnaires afterwards. As an ethnographer I was faced with adapting my proposed methodology to the varied challenges I faced in the field. Indeed, it is a common occurrence that 'the realities of the field situation may dramatically alter months of careful planning' (Crick 1989: 25). Despite these challenges, by both co-ordinating the Oxfam study tours and participating and observing the tour participants' experiences I was able to obtain a level of detail that would help to inform my broader objectives of critically investigating the development-tourism nexus.

As there are many ways of undertaking lines of inquiry, my first and principal method was to be active and passive participant observer on NGO study tours. To examine study tours and what the participants experience while in Cuba involved dealing with certain difficulties and cultural sensitivities. An example was preparing people for the machismo that most tour participants are not familiar with in their own societies. Not only did I continually observe the informal interactions and dialogues between tour participants, I also recorded the seminars that were mostly in Spanish and translated by the Cuban tour leaders. I interviewed the advisor to the Minister of Tourism and collected literature and documents regarding tourism and development in Cuba. My observations were not limited to viewing the tour participants on tour but included the analysis of a variety of texts from the field, for example recording of presentations to the tour participants by government, non-government, development and grassroots organisations and subsequent dialogue between the group and presenter, and observing the interactions between foreigners and locals at development projects. These wider inquiries provided a broader context within which to situate the findings of the questionnaires.

The implementation of post-tour questionnaires is a reasonably non intrusive research technique where respondents can answer in their own time rather than under the pressure of a face-to-face interview. In fact one might argue that data from such questionnaires can be equally if not more illuminating than face-to-face interview data. For this to be possible, questionnaires must be designed and analysed not only as a positivist tool eliciting facts and 'literal' measurement of attitudes. They also need to be viewed as a discourse containing constructionism, reflexivity and narrative structure. The questionnaires, not only attempted to elicit data but also to document the naturalistic discourse that their experience produces. In this way, the research participants were the participant observers as much as myself in that they were documenting their perceptions of their experiences of learning about development in Cuba.

Initially I intended to interview study tour participants either during the tour or immediately post-tour while they were still in Cuba and I anticipated that most of them would be keen to discuss their experiences. However, the logistics involved in undertaking many interviews proved inappropriate for this research setting; another problem with researching tourism. I hadn't anticipated that there would simply not be time available in the itineraries to devote to in-depth interviewing of numerous participants *and* that most participants immediately left the country as they had to get back to their work/study/family commitments back home. I was able to explore in more detail the impacts of tourists' experiences on their long term attitudes and behaviours and hence what implicit consequences these might have for development.

Book Structure

The book is structured in such a way as to guide the reader through three parts each having two chapters. The first part – Chapters 1 and 2 – present the theoretical discussions that inform the research by providing critical perspectives on major tropes underpinning development, globalisation and tourism. Part 2 – Chapters 3 and 4 – presents the context of the study by providing the reader with background to Cuba, its development paradigm, and an account of NGO study tours. Part 3 – Chapters 5 and 6 – comprise an ethnography of the tourists who participate in NGO study tours to experience encounters with development.

Chapter 1 situates the study within the context of development, especially literature concerning how development bodies are beginning to embrace tourism in varying ways within policy and practice. The chapter considers why development policies have recently embraced tourism and to what extent development can utilise increasing global flows of people. These questions highlight the centrality of development and globalisation for an in-depth understanding of the NGO study tours taking place in Cuba. The chapter highlights how globalisation debates inform dialogues about development particularly as it pertains to international tourism and its increasingly inferred development potential. The chapter draws attention to the corresponding changes in studies, practices and related institutional policies pertaining to development and tourism and how these two domains have followed a similar path in developing countries. The discursive impacts of these shifting approaches are looked at to shed light on the emerging moral imperatives in the development process, namely those of sustainability, pro-poor growth and rights-based approaches. According to many, development has reached a point where it needs rethinking. In view of that, tourism is increasingly incorporated into development initiatives.[6] Cuba, being a socialist country, provides a unique

---

6   Department for International Development operate a scheme called pro-poor tourism aimed at helping the very poor in rural areas (see Chapter 1). USAID use ecotourism as a strategic development tool. SNV (Netherlands Development Organisation) deploys

perspective on the extent to which this model of capitalism and tourism are natural allies in the attempt to improve the economy and livelihoods.

The second chapter embeds the research within the emerging literature on the shifting nature of tourism to provide a background to the specifics I consider in following chapters. In this chapter I extend the arguments of Mowforth and Munt (1998; 2003) about tourism and globalisation and Poon's thesis on 'new tourism' and the new middle classes. I review the key discourses that underpin and inform tourism demonstrating the ways in which sustainability, tourism, globalisation, and development intersect. I am particularly concerned with how notions of the moral infuse 'new tourism'. I outline the values and characteristics associated with 'new tourism' and discuss how sustainable, responsible and appropriate forms of development are dominant themes implicit in new forms of tourism which claim a moral superiority to mass tourism. Moving beyond Butcher's (2003) critique of tourism I take his notion of 'the moralisation of tourism' in a different direction to arrive at a new analytical outcome that merges morality and ethics with notions of rights-based development. Departing from Butcher's critique I begin to look at how morality is not simply about superiority but also about ideas of responsibility and solidarity. This is illustrated by a case study of the two organisations that facilitated my research in Cuba. The NGO study tours represent examples of new moral tourism in practice, demonstrating one of the ways in which development and tourism intersect. This discussion lays the groundwork for the specifics I consider in the chapters that follow.

Part 2 of the book provides the context of the study by drawing on my ethnographic material about social development in Cuba and introducing the reader to the concept of NGO study tours. Chapter 3 introduces the site of my research – Cuba. It details Cuba's model of development, the US-imposed embargo and its effects, the 'Special Period', and the subsequent economic reforms and social changes. It is important to understand Cuba's economic crisis in order to appreciate the nature of its social development and participation by local people.

Chapter 4 immerses the reader in the tourist-development encounter in Cuba by presenting some excerpts from my field notes on development-oriented tours in order to give the reader a sense of what it is like to participate in a NGO study tour in Cuba and in what ways the participants learn about Cuba and its development practices. The chapter highlights the messages that Cuban people and grassroots organisations want tourists to assimilate and how development-oriented tours are a means of advancing Cuban solidarity in an increasingly globalising world. The people and organisations in Cuba that tour participants meet promote solidarity. Organisations such as the *Cuban Institute for friendship of the People* (ICAP) proclaim tour participants as "ambassadors" for the Cuban cause. For the tour

---

sustainable tourism as a poverty reduction tool. The Green Travel Guide argues for tourism as a conservation tool. And NGOs are promoting tourism for awareness of development, human rights and conservation issues, i.e. Global Exchange, Earthwatch, WWF, Audubon Society, and more.

participants themselves the study tours are an opportunity to support community development: through the money they pay for their tour, through the opportunities to learn about development, and to travel with other like-minded people.

Part 3 of the book provides an anthropological exposition of tourist motivations and experiences. Drawing extensively on my ethnographic material Chapter 5 examines the motivations of people who participate in NGO study tours in Cuba framing the analysis in personal transition literature of anthropologists. We see that what motivates people to participate in the tours is multifaceted. I investigate whether they are driven simply by a desire for cultural capital gain, or if their interests go beyond this to have more meaningful results for development? This provides the material with which to question notions of rights-based tourism and its functional implications. We learn that increasingly people are motivated by a desire to spend their money in ways that will support development while at the same time seeking opportunities to learn about issues which affect people around the world. This feeds into broader tropes of sustainability and rights-based development in which people make decisions based on moral imperatives and this engenders solidarity. The final ethnographic chapter demonstrates how people participating in NGO study tours fit within a new moral tourism paradigm based on their experiences, which often leads to an explicit or implicit personal transformation. This chapter demonstrates that issues of morality underpin the merging of this form of tourism in Cuba with rights-based notions of development. It is this convergence between tourism and development that leads to positive outcomes like the creation of network ties, participation in new social movements, and solidarity.

PART 1
Critical Perspectives Underpinning Development and Tourism

PART I

Chapter 1
Development and the Rise of Tourism as a Strategy

The interconnections between globalisation, development and tourism are a crucial nexus in analyses of North-to-South exchange and in relation to 'empire' (c.f. Hardt and Negri (2000); Nederveen Pieterse (2004) in relation to arguments for and against purported empire). The recent shifts in paradigmatic approaches to the alleviation of poverty throughout the Third World provide a rationale for the move to incorporate tourism strategies within development. By highlighting the shifting perspectives of development theory and practice, I demonstrate that the development industry's history of adoption and casting off of new strategies leads us to question the viability of tourism as sustainable development. In fact it alerts us to the likelihood that tourism may only be a temporary strategy within the development arena. I include this exposition of the shifts in development modes and praxis to embed my research of tourism within a development context. This is vital because the study tours that are the focus of this research are operated by NGOs as part of their programmatic initiatives and the major intention of the tours is to expose participants to (and thereby assist in) Cuban development.

Just as perspectives in development studies have changed over time from economic to social, to environmental, to ethical, so too have perspectives in tourism studies, with the emphasis on economic factors changing to include sociocultural and environmental perspectives. Thus changes in development and tourism studies bring to light the parallel trajectories that discussions on development and tourism in Third World countries have followed. What I find particularly intriguing is the similarity of moral imperatives driving more recent trends in social change initiatives such as sustainable, pro-poor and rights-based development with those that arise within tourism.[1] As with tourism, development debates fuse compellingly with those on globalisation, particularly globalisation as cultural flows and neoliberal economic policy. We can begin to understand the emergence of debates about global tourism and its inferred development potential providing a background for the focus on 'new' tourism and its associated moral underpinnings discussed in Chapter 2.

---

1 A sense of moral duty has always underpinned Western development interventions but since the Millennium Development Goals it is now arguably more pronounced as an imperative of all aid assistance.

Globalisation: Reality and Myth

Recent shifts in development thinking stem from the prevailing neoliberal ideology guiding the world economy, where we see that many things have globalised but not wealth and development in any pervasive sense. Current debates around globalisation dovetail with theoretical disputes concerning the ongoing pursuit of development and modernisation. As Sachs concludes, since the 1980s the age of development "has given way to the age of globalisation" (1999: xii). As with capitalism and modernity, globalisation is a "megatrope" (Knauft 2002: 34). Globalisation, of the neoliberal economic policy kind, has become increasingly dominant since the 1980s but there are other prevailing meanings of globalisation; the increased integration into a world economy and the effects of improved communication and transportation systems on multidirectonal cultural flows. Theories of globalisation thus inform debates about global tourism and its development potential from the 1980s onwards.

Some scholars suggest that both development and globalisation are nothing more than myths that have been constructed by Western exploitative economic interests to promote Western democracy as political modernisation (Crush 1995; Kothari, Minogue and DeJong 2002). Until the 1980s development aid was a major tool of the Cold War and with the breakdown of the Soviet Empire, the presumption is now that there are no obstacles to global modernisation. We might question then how globalisation differs from previous periods of global interconnectedness, because parallels between modernisation and globalisation can easily be drawn over various historical periods. Bourdieu (2001: 2), for example, says globalisation is:

> a simultaneously descriptive and prescriptive pseudo-concept that has taken the place of the word 'modernisation', long used by American social science as a euphemistic way of imposing a naively ethnocentric evolutionary model that permits the classification of different societies according to their distance from the most economically advanced society, which is to say American society ...

For many anthropologists, to invoke the global is to highlight the speed and density of interconnections between people and places (Tsing 2000), as well as disconnections, exclusion, marginalisation, and dispossession (Appadurai 1996, Ferguson 1999). It is argued that the key difference from previous global interconnectedness is located in the speed by which global forces penetrate across cultures (Harvey 1989). There is far more interaction throughout the world over a shorter timeframe, resulting in immediate consequences in one region from the actions in another. Giddens (1989: 520) states that "Our lives ... are increasingly influenced by activities and events happening well away from the social context in which we carry on our day-to-day activities". The emergence of international institutions since the 1940s such as the UN, the World Bank, the IMF, transnational corporations and the more recent growth of non-governmental organisations

(NGOs) and the global reach of their policies is indicative of this (Hardt and Negri 2000). Technological advances, economic and political shifts, cultural change, increased communication and travel, are driving forces for globalisation and are all closely linked. The so-called 'Washington Consensus' on global deregulation of financial markets, post-Fordist production and the proliferation of transnational corporations led to economic transformations and rapid growth in financial flows. At the same time, and closely linked to economic change, we saw political shifts in the decline of State intervention and the emphasis on deregulation, privatisation and liberalisation. These changes occurring within the climate of globalisation have inspired a rethinking of the role of the nation state and particularly its involvement in development (Kothari and Minogue 2002: 18-19). It has also meant that the profits generated from increased flows of international tourists into Third World countries are flowing offshore to transnational corporations.

In addition, and some might say, as a countervailing force, there has been an increase in international social movements and increasing awareness of global issues. Throughout the 1980s, social movements (environmentalism, human rights, indigenous rights, and feminist causes) established themselves through NGOs and moved beyond the boundaries of nation states by producing transnational avenues of financial and political support (Tsing 2000). This has had the effect of uniting people beyond national borders on issues such as the environment, human rights, war, refugees and resettlement. Additionally, advances in technology have led to increases in the speed, intensity and quantity of global flows of information, ideas, capital, goods and people, which raise questions about the uneven distribution of the benefits of globalisation. In the context of tourism in Cuba this has meant that the increased mobility of international tourists is in stark contrast to the immobility of the majority of the Cuban population.

Appadurai (1990) suggests that we can understand globalisation as a series of global disjunctive flows that create different 'scapes'. For example financescapes include capital flows and ethnoscapes include the cultural worlds conjured by migrants. In this way, he refers to the paths taken by those things that are in flow as:

> hav[ing] different speeds, axes, points of origin and termination, and varied relationships to institutional structures in different regions, nations or societies. Further, these disjunctures themselves precipitate various kinds of problems and frictions in different local situations. Indeed it is the disjunctures between the various vectors characterising this world-in-motion that produce fundamental problems of livelihood, equity, suffering, justice and governance. (Appadurai 2001: 6)

For example the media flows across national borders that produce images of well-being unrealisable by national living standards are examples of the disjuncture referred to by Appadurai. We might also consider tourist flows across borders that produce notions of wealth, status and freedom unrealisable by many people in

Third World countries as yet another example. These disjunctures highlight that globalisation produces uneven benefits and many locally manifested problems (Appadurai 2001: 6). As has been the case with development, globalisation needs to be interrogated.

Neoliberal Orthodoxy:
Intersections between Development, Globalisation and Tourism

The relationship between developing countries and tourism is dynamic and framed within broad global flows and social changes. In order to demonstrate in what manner tourism has increased in developing countries, discussions of tourism invoke the term globalisation and refer to a global community, a continually shrinking world in which countries are increasingly interdependent. Globalisation is hardly a new phenomenon – capitalist development has historically been global by nature. What differs today is the rapid nature of global processes and change. It is often argued that tourism contributes to the creation of globalising cultural forms that tend to erode cultural differences. Critics of this view dismiss the idea of an impending global homogenous culture. Tsing (2000: 39) states that "no anthropologist I know argues that the global future will be culturally homogenous; even those anthropologists most wedded to the idea of a new global era imagine this era as characterised by 'local' cultural diversity". MacMichael (1996) refers pointedly to Western developmentalism rather than development, stating that it more accurately indicates the nature of much development ideology which is based on economic theories of technology and modernisation. He argues further that developing countries should be given opportunities to develop in ways that respect local culture and are more in keeping with local culture. Indeed, Butcher (2003: 98) points out that the view that cultural difference is generally ignored in development has resonance within tourism where such perspectives are also commonplace.

   Globalisation provides a framework to apprehend the cultural, economic, and political aspects of global change (Allen and Massey 1995; Mowforth and Munt 2003). Simply, cultural globalisation refers to a trend toward homogenisation of culture, a single global monoculture, brought about essentially by global consumerism. As we shall see, the critical school of tourism theorists sees mass tourism as the inauthentic consumption of places and cultures, which tends to reduce local people, traditional cultures and indigenous knowledge to commodities. Such simplistic views of tourism need to be problematised within broader understandings of how local people in developing countries harness tourism to their own advantage, for example, utilising pro-poor or rights-based development models as a framework for tourism and a topic I shall return to.

   Economic globalisation refers to the way fiscal processes now cover the world. The rapidly expanding international tourism industry increasingly incorporates developing countries as 'new' destinations. At the same time, the boom in other industries such as telecommunications and transport has facilitated the globalisation of tourism. However, globalisation is not just indicative of economic change, but also of a politically changed world that some scholars argue reinvokes empire while others suggest contemporary globalisation falls short of empire (c.f. Hardt and Negri 2000; Ferguson 2002; Nederveen Pieterse 2004). The reach of large-scale international organisations such as the European Union, International Monetary Fund (IMF), World Bank (WB), World Trade Organisation and international environmental organisations like Greenpeace evidence this politicisation of supranational organisations. The work of these organisations is felt all over the world and impacts differently on different groups of people, highlighting the power relations that underscore uneven development. Domination is now exercised through financial and economic regimes. US-imposed sanctions on Cuba and Iraq evidence this pattern. Political globalisation typically centres on the loss of sovereignty of nation states through increasing transglobal organisations and politics and the consequent breaking down of national borders (Mowforth and Munt 2003).

The post WWII emergence of supranational institutions and agencies has had a significant impact on global development and on international tourism. International organisations like the WB and IMF have had pervasive effects on the development of Third World countries by imposing structural adjustment policies which force them to adjust their economies in order to secure further loans. WB and IMF have become more involved in tourism mainly because, even when commodities are at low levels or agriculture is eroded by subsidised competition or services are non-existent, developing countries, can and do, offer venues for tourism. The WB and IMF perceive that given input from the development community, tourism is an area where developing countries have a comparative advantage and could operate sustainably and profitably. This can be seen through numerous initiatives like the World Tourism Organisation signing an agreement with the WB and accords with the regional Development Banks so that they place tourism higher on their agendas, and the World Tourism Organisation supporting new research and funding thousands of new micro projects by 2015 worth hundreds of millions of dollars as a contribution to the Millennium Development Goals. In addition, the World Tourism Organisation in its capacity as the new UN tourism organisation has focused on increasing trade liberalisation and is developing with the World Trade Organisation and UNCTAD[2] a number of initiatives to enhance tourism exports as a development tool (Lipman 2004).

Other intergovernmental structures that transcend the autonomy of the nation state include NAFTA[3] and ASEAN.[4] The political relationships forged by these institutions have extended their reach globally and have adopted the development language of sustainability. At the same time there has been an emergence of

---

2 UN Conference on Trade and Development.
3 North American Free Trade Agreement.
4 Association of Southeast Asian Nations.

international NGOs. Through this global political environment, we see that sustainability does not simply pertain to the environmental context. Sustainability refers to ongoing profits through adaptable patterns of capital accumulation and even the survival of indigenous cultures which Western middle classes experience when travelling to developing countries.

The global nature of tourism inevitably dovetails with processes of modernisation, industrialisation, economic development and the complex nature of social change in this period of global interconnectedness. As mentioned, the debate on globalisation intersects with discussions on development, particularly in relation to the uneven and unequal nature of both development practice and globalisation. In the context of tourism, an example that highlights global interconnectedness and how countries have become interdependent is the October 2002 bombing on the island of Bali in Indonesia. Bali thrives on international tourists. When a night club on the island was the target of a political religious attack, killing many Balinese and Australians, the Foreign Affairs Department in Australia advised Australian tourists to avoid travelling to Bali and this advice was immediately copied by other countries. The Balinese tourism industry came to a grinding halt and subsequently the local economy almost collapsed. This example illustrates impacts of globalisation in that Bali, a developing island economy dependent on its international tourism industry, is more adversely affected by the cessation of tourism to the island than are its Western tourists.

In discussions of global interconnectedness, most accounts of globalisation are Western as a result of the expansion of capitalism – the export of Western goods, ideas, values, and people. Arguably globalisation thus represents Western ethnocentricity. It is claimed by some that globalisation has helped First World governments and businesses to determine a Western inevitability that incorporates a global economy, culture, politics and environment (Mowforth and Munt 1998). Examining globalisation closely can help in our understanding of issues of sustainability. As with development, sustainability is uneven in nature in developing countries as concern for the environment has resulted in a 'green' politics. From a tourism perspective Mowforth and Munt (1998) state that it is necessary to trace how debates about the environment, development and sustainability relate to 'new' forms of tourism. The first important point they make is that sustainability is defined, interpreted and practised differently according to the interests of different groups. Sustainability is a contested concept that is socially constructed around competing agendas. Secondly, any discussion of sustainability should consider the different ways in which it is practised, how sustainability is appropriated by social classes as a means to represent identity, and how local communities use policies of sustainability to exclude tourists (Mowforth and Munt 1998).

This is relevant to NGO tours in Cuba for two reasons: first because development-oriented NGO tourism offers small group tours to development projects and grassroots organisations in an effort to be small scale and therefore easy to sustain in that they do not impact largely on small communities being visited. Second, participation in what are considered to be sustainable forms of tourism feed into a

Western middle class identity. Such tourism is emerging in Cuba as its depressed planned economy has been reinserted into the capitalist world market. Throughout the 1990s elements of capitalism have been reintroduced and concessions made to market mechanisms, such as foreign investment, a temporary legalisation of the US dollar, a free market in agricultural produce, the expansion of self employment, and a tourism led recovery. Thus the sustainability of tourism as a development tool in Third World countries emerges precisely because development practices have proved to be unsustainable and for Cuba in particular this is a crucial point in time as it strives towards reinsertion into a global economy. Hence, tracing the trajectories that have led to an impasse within development is important because it shows us that national governments and international institutions have lost much of their legitimacy and non-governmental organisations such as Oxfam and Global Exchange emerge from and operate within this context.

An Intellectual Heritage of Development

The continued existence, and in some cases the exacerbation of, poverty in developing countries juxtaposed with the immense volume of research and funding being undertaken to counteract these problems indicates that institutionalised development programs intended to improve livelihoods have been fraught with shortcomings. Likewise, theory and attempts to represent the meaning of development remain enormously vexed. Critical interpretations are continually rethought while notions of an apparent 'impasse' in the framing of theoretical questions reflect the very contentious nature of determining the best way to 'do development'. At the same time, neoliberal imperatives continue to drive forward new waves of capitalist modernisation. The current move in development thinking towards the use of tourism as a tool to help reduce poverty can be seen as one recent manifestation of shifts in aid models and leads us to the very basic question of why development policies have recently embraced tourism. A partial answer indicates that as a tool, tourism intersects with models of development that have come before it – sustainable, participatory, pro-poor and rights-based – in productive but not unproblematic ways.

Importantly, this move to enlist tourism signals that development in recent decades has been deeply entwined in both conceptual and material aspects of globalisation, particularly through neoliberal economic policies, the increasing integration of developing countries into the world economy, and the effect on multidirectional cultural flows of improved communication and transportation systems. This then raises a second question: to what extent can development utilise these global flows of tourism? These lines of enquiry illustrate the centrality of development and globalisation to an in-depth understanding of the particular tourism forms and experiences taking place in Cuba. In a position to learn from past mistakes in international tourism development, Cuba claims to support sustainable forms of tourism. In doing so, the Cuban government endeavours to keep the tourism economy separate from the national economy in an attempt to curb foreign leakage, and importantly, the corrosion of socialist values. Mostly, tourism development in Cuba has taken the form of enclavic resorts. But there are also niche participatory forms of tourism that have positive, solidarity, and development aspects.[5] And there are niche forms of tourism based on working with local communities, for example the Friendship Brigades who stay and work with local families (usually in agricultural activities). In addition there are growing numbers of NGO study tours to meet local people and learn about social development issues. My book focuses on these tours, but before I address the specifics of them, I will trace the intellectual heritage of development in order to contextualise the emergence of this tourism-development conjunction taking place in Cuba.

Notions of development can be traced back to the late eighteenth and nineteenth century rise of industrial capitalism in Europe which addressed the alleviation of unemployment, migration and poverty. Development as a body of thought and a series of programmatic initiatives has taken different forms since its inception with highly divergent interests. Thus theories which guide development processes have undergone multiple paradigmatic shifts from economistic measures such as GDP growth, to visions of desired social change, in particular improved well-being.

The development teleologies of Enlightenment, Marxian, Weberian and other notions of progress have centred on economic growth. The establishment of the Bretton Woods financial institutions – the World Bank and International Monetary Fund – occurred in 1944. John Maynard Keynes, foundational in the establishment of these international bodies emphasised economic policies such as fixed currency exchange rates and the institutionalising of national economic planning to promote growth (Edelman and Haugerud 2004). This approach to development, centred on State sovereignty and thereby encouraged State intervention in the economy.

Development became a specific international initiative to reduce world poverty after US President Harry Truman's inauguration speech in 1949. The world superpowers began to refer to poor countries as the 'Third World' and 'development' as the economic growth process that they should follow in order to move up the ladder of modernisation, with technology and capital the most potent symbols of modernity. The sixty years since the (in)famous Truman speech proclaiming the southern hemisphere to be 'underdeveloped' have been defined as the 'age of development'. The idea was that countries in the South reproduce the development processes of Europe and the United States. Truman's appropriation of the word 'underdevelopment' symbolised the beginning of an era of American hegemony (Esteva 2001). In this context, the word took on a powerful colonising virulence. According to its more trenchant critics, the idea and practice of development gave global hegemony to a purely Western genealogy of history, robbing people of different cultures of the opportunity to define their social life.

---

5 Daniel, Y. (1995) *Rumba: Dance and Social Change in Contemporary Cuba.* Indiana University Press: Bloomington.

In contrast to many Latin American countries, Cuba did not subscribe to the borrowing of capital from the Bretton Woods institutions. Its economic policies were virtually the direct opposite of the Washington Consensus. The government controlled practically the entire economy, permitting private entrepreneurs in limited industries. It heavily subsidised virtually all staples and commodities; and its currency was not convertible for a long time. It retains tight control over all foreign investment, and often changes the rules for political reasons. At the same time, however, its record of social achievement has been sustained and enhanced. This is discussed in Part 2 of this book.

Definitions of social development were vague throughout the world in institutional circles until 1962 when the Economic and Social Council of the United Nations finally acknowledged the importance of incorporating both the economic and social aspects of development and the following year saw the creation of the United Nations Research Institute for Social Development. This period became characterised by competing theories of modernisation and dependency; divergent bodies of thought that depict development respectively as positive change or negative domination. The construction of large resorts in the Caribbean, for example Cancun in Mexico and Varadero in Cuba, were designed as catalysts of modernising economic growth with the trickledown effect in mind. However, as development specialists have learnt, defining development as economic growth is problematic because in reality trickle down rarely occurs (Gardner and Lewis 1996), allowing criticism to suggest it create situations of exploitation rather than assistance.

By the 1970s, modernisation theories failed to explain why most Latin American countries remained poverty stricken, despite nearly a century of independence. Perhaps encouraged by socialist advances without communist government, this Latin American problem inspired new development thinking best known as the neo-Marxist dependency or underdevelopment theories. This focused on how historical processes had exploited and stripped the former colonies of their resources in order to aid the industrial revolution in Western countries (Sylvester 1999). Theorists of this ilk, '*dependentistas*', see lack of development in Third World countries as a result of power at the centre exploiting an underprivileged periphery as was the case with colonialism (c.f. Andre Gunder Frank 1967; Immanuel Wallerstein 1974). Such relationships have been viewed by some scholars (c.f. Britton 1982, 1989; Nash 1989) as being perpetuated by international tourism, arguing that tourism also demonstrates neocolonial tendencies. World systems theory usefully provided a means of differentiating between internal and external factors as explanations for underdevelopment, whereby underdevelopment stems from the lack of incorporation of countries into the world system because countries are subject to unequal trade, *not* because of internal factors such as traditional values and culture inhibiting social change. In the context of tourism development, dependency theorists have rejected the involvement of multinational corporations (like Western owned Club Med style resorts) and advocated small scale locally owned tourism ventures.

The economic focus became more prominent throughout the 1980s and 1990s as "a response to the neoliberal intellectual climate where multilateral and bilateral donors gave the impression that hard-headed, monetary cost-benefit analysis in planning and evaluation documents was indicative of professionalism" (Crewe and Harrison 1998: 38). The recent policy move to embrace tourism by various development institutions can be interpreted in this light as a move away from the notion of technology as *the* catalyst to development. The WB, the UNDP, the Asian Development Bank and others have recently helped fund significant tourism programs throughout the developing world as an indication of refocusing their projects on local participation.

The merging of the economic and social aspects of development coincided with the breakdown of the Bretton Woods controls on capital movements. Throughout the 1980s and 1990s the WB and IMF advocated a radical set of reforms referred to as structural adjustment policies to guide the development process. This represented a huge shift in development thinking from their early policies. Where these programs had previously emphasised the role of the State, they now sought to reduce the role of the State in the economy and introduce structural adjustment policies of reduced State expenditure on social services, user pay fees for social services, currency devaluation, selling of State owned enterprises, and deregulation of financial markets. Structural adjustment had exacerbated Third World debt and plunged more people into poverty. The former WB vice-president Joseph Stiglitz (2002) strongly criticised the impact of structural adjustment policies on the living standards of the poor.

The impact of these policies has prompted specialists to suggest that development has not only failed in its undertakings but is inherently flawed and thereby simply perpetuates poverty. In his book *The Myth of Development*, echoing many other theorists, Oswaldo de Rivero (2001: 114) claims that:

> International aid, the daughter of the myth of development, is paradoxically the clearest testimony of non-development. During nearly half a century, the United Nations, industrialised powers, specialised agencies, international financial organisations, non-governmental organisations and humanitarian institutions have tried uncounted policies, strategies, programmes, and development projects, transferring billions of dollars in credits, technical assistance, equipment and donations ... and only a modest stream of this torrent of resources has been applied to alleviate poverty.

De Rivero's book represents a move by some scholars to question and reject Western intervention. Following Foucauldian lines of enquiry, some suggest Western discourses about development are deeply implicated in practice through the imposition of ideas of modernity on non-Western societies. Discourse is one of the ways through which power operates, as the knowledge which a discourse produces implicitly constitutes power (Hall 2002: 63). Drawing on Foucault some claim (c.f. Escobar 1995) that through its discursive impact, development has manufactured categories and given a new value system associated with these that was inevitably internalised within local subjectivities. Categories which development could treat and reform such as 'the poor', 'underdevelopment', and 'Third World'; they thus see development as a self serving process engendered by bureaucrats and development professionals that permanently entraps people in Third World countries in a vicious cycle of poverty. These scholars are proponents of 'post-development', which embraces 'indigenous' knowledge and new social movements as a means of directing alternatives to aid assistance (Esteva 1988; Escobar 1995; Rahnema 1997; Sachs 1992; Illich 2001; Hall 2002). In contrast, as less radical responses, others have emphasised alternatives *within* existing development thinking (Crewe and Harrison 2002; Little and Painter 1995; Nolan 2002).

Such theoretical positions raise the question whether in fact poverty is strategically deployed within the development-tourism nexus, for example, the recent development strategy of pro-poor growth that incorporates tourism and which I will come back to later in this chapter. For Escobar (1995), the most important exclusion from development has always been what development was supposed to be all about – people. "Development [is] a top-down, ethnocentric, and technocratic approach, which treat[s] people and cultures as abstract concepts, statistical figures to be moved up and down in the charts of 'progress'" (Escobar 1995: 44). This perspective denies the benefits that the international development institutions have brought to some people around the world and we see many examples where his notions are not entirely accurate. NGO study tours, as a more recent example, bring people to the forefront. This is precisely because their focus is to meet with local people to discuss and learn about and exchange development knowledge.

Having undergone a turbulent history, development for many intellectuals "has become a non-word, to be used only with the inverted commas of the deconstructed 1990s" (Gardner and Lewis 1996: 1). However, mobilisation around the Millennium Development Goals (MDGs) by many of the world's governments demonstrates that it is undoubtedly useful to consider options beyond rhetoric that engage the multiplicity of voices of those impacted by development. In so doing, there is a pressing moral responsibility to work towards improving the quality of life for the large majority of the world, despite the fact that most countries are struggling to stay on target for most of the MDGs.

Moral Imperatives in the Development Process

Three development trends that have emerged in the new millennium as dominant discourses driven by a moral and ethical underpinning include sustainability, pro-poor growth and rights-based development. These recent trends in development approaches intersect at different discursive and practical levels which supposedly give a moral integrity to international programs of assistance and as we shall see are driving new forms of alternative tourism.

Sustainable Development

Increasingly, development paradigms have become more complex as an emphasis on economics has given way to more holistic perspectives incorporating a wider range of concerns. Sustainable development has been defined by the World Commission for Environment and Development (1987: 8) as development "that meets the needs of the present without compromising the ability of future generations to meet their own needs". The concept of sustainability emerged from environmental concerns and embraced within the alternative development paradigm that advocates meeting basic needs of food, shelter, water, health and education. The contested and political nature of development illustrates that notions of sustainability emerged into a field heavily debated and with strong ideological underpinnings. Arguably, sustainable development is a concept that recognises the needs of poor people and the limitations imposed on development by current levels of technology, social organisation and environmental variability (Wood 1993: 7). It has received strong bureaucratic support from the whole arena from grassroots organisations to international agencies; partly because it creates opportunities at all levels (institutional, policy, programmatic) to expand power bases, acquire additional resources and enhance prestige (Wood 1993: 7). Therefore, key questions to be considered are "what is being sustained, by whom and for whom; do all interest groups have the same intentions or aspirations in terms of sustainability? ... who decides what sustainability means and entails, and who dictates how it should be achieved and evaluated?" (Mowforth and Munt 1998: 12).

Within sustainable development a grassroots perspective is proposed which emphasises public participation and local planning. Unfortunately, the focus of many governments in developing countries is not predominantly on the poor but on large-scale projects especially in tourism (such as airports) which provides little opportunity for local participation. Thus the concept of sustainable development, especially in tourism, has often meant a gap between policy and practice. Sustainable forms of tourism development have been labelled ecotourism, nature tourism, appropriate tourism, ethical tourism and responsible tourism, all of their meanings blurred and overlapping, but all broadly fitting under the banner of sustainable tourism.

In understanding the complexities of tourism as a sustainable developmental tool, it is important to recognise that, inevitably, tourism takes place in the context of great inequality of power of the sort described by Escobar. In this way some sense of moral adjudication has become significant in a climate where damage to culture and environment is now a prominent issue. It is thus necessary to consider how hegemonic discourses and ideologies impact on tourism in developing countries. In fact, development, sustainability and tourism are all examples of hegemony in practice (Mowforth and Munt 1998). Tourism is laden with highly publicised strategies including tourism codes of conduct and the advocacy of more responsible and sustainable forms of tourism.[6] The term sustainability has been appropriated to give moral integrity and green credentials to tourist activities. Such initiatives indicate a recent move in the tourism industry towards modifying practices for a more sustainable tourism that is guided by ethical underpinnings. This grows out of larger tropes of postcolonial studies and rights-based development concerns over subaltern subjects allowed freedom from prior modes of oppression. Development, it is often argued, has the capacity to impinge on peoples' rights. Tourism strategies are now considering tourism's role in power relations and the extent to which it is either improving or degrading people's lives. In this way development and tourism are intersecting a number of tropes at different discursive and practical levels.

Hence in this era of globalised interconnectedness, uneven development and unequal relations of power, the burgeoning international tourism industry is being positioned as a key player in sustainable development as evidenced by Agenda 21 – the international community's response to the need for devising strategies to halt and reverse the unwarranted effects of poverty and environmental degradation and which designates national governments as being responsible for developing sustainable tourism. To this end, poverty alleviation has been drawn into the sustainability discourse. Pro-poor growth is one such strategy which looks specifically at how tourism development can contribute to a more directed approach to ensure that benefits from tourism flow to the poorest sectors of communities. Aid donors and international funding agencies increasingly proclaim tourism to be an effective tool for poverty alleviation and state that the poorest segments of populations need to be empowered if alleviation of their plight through tourism is to be sustainable. Other agencies have developed sustainable tourism initiatives with a poverty focus, such as the World Tourism Organisation and UNCTAD's program 'Sustainable Tourism for Eliminating Poverty' (STEP) and the CRC Sustainable Tourism's (Australia) 'Sustainable Tourism Actively Reducing Poverty' (STARP).

Pro-poor Development

While development remains contentious both theoretically and as experienced by those under its programmatic impact, one way or another, poverty has resurfaced as a key element of evolving approaches. The renewed focus on poverty reduction as the fundamental goal of development, as evidenced at the summits in Doha (trade) 2001, Monterrey (debt) 2002, and Johannesburg (sustainability) 2002, and epitomised by the Millennium Declaration, has engendered great interest in the

---

6 WTO's 'Global Code of Ethics for Tourism'; Tourism Concern's 'Code for Responsible Travellers'.

concept of pro-poor growth[7] and is argued by many to be the most significant policy measure to achieve this goal. This return to poverty reduction in development thinking reflects the lack of success in development initiatives to date, not the least because poverty still pervades the globe. Tourism is currently being tested as one source of such growth and this is the key to appreciating the complex underpinnings to the orchestration of tours to Cuba. Pro-poor growth, albeit a continuation of the conventional thinking that has informed development for decades, as it is an economic based strategy focusing on the income dimension of poverty, has a distinct moral complexion. It is an emerging development strategy that is driven by a moral concern for reducing poverty above all else. Discursively, 'pro-poor' indicates that previously development has not centred on poor people or that it has in hindsight created a growth in poverty.

Klasen (2001) asserts that most of the many current definitions of pro-poor growth are imprecise and inconsistent, implying that it refers to growth that leads to significant poverty reductions (United Nations 2000; World Bank 2000). Further, he argues there is still much ambiguity surrounding what the policy implications of a call for pro-poor growth are. He thus attempts to propose a precise definition whereby: "the poor benefit disproportionately from economic growth. This is to say the proportional income growth of the poor must exceed the average income growth rate" (Klasen 2001: 2).

More broadly, pro-poor growth is defined as that which allows poor people to actively participate in and benefit from economic activity. It theoretically represents a significant departure from the trickle-down development concept as economic growth forms only part of the development process along with social factors such as everyone having access to minimum basic capabilities. Kakwani and Pernia (2000) argue that implementing pro-poor growth strategies therefore needs to be specifically aimed towards the poor in order for the less resourced sections of society to benefit proportionally more than the rich. In addition, these strategies require eliminating institutional biases against the poor and adopting direct pro-poor policies such as sufficient education, healthcare, improved access to credit, and the promotion of small and medium enterprises (Roe and Urquhart 2001).

Arguably, tourism can be utilised as one source of pro-poor growth. Pro-poor tourism development may focus particularly on economic aspects but also incorporates aspects of the sustainability and participatory models of development, thereby elevating the moral imperative of this style of tourism by ensuring that local interests are met. A joint report undertaken by the Overseas Development Institute (ODI), the International Institute for Environment and Development (IIED) and the Centre for Responsible Tourism (CRT) argues that a co-ordinated strategic approach to making tourism work for the poor is highly important. Contemporary development thinking on poverty reduction emphasises the complexity of the

---

7   For comprehensive discussions on the relationship between economic growth and poverty reduction see McKay 1997; Ravallion 1997; World Bank 2000 (chapters 3-5).

process and the need for a variety of complementary strategies. Tourism in all of its forms is one of the world's largest economic sectors and therefore ODI, CRT and IIED argue it can make significant contributions to economic growth and poverty reduction. Pro-poor tourism is not a particular product or even sector of tourism but rather an approach to tourism that involves a variety of stakeholders including government, private sector, civil society and poor people who operate both as producers and decision-makers (Ashley et al. 2001).

Tourism affects the livelihoods of many poor people around the world. However, tourism's significance clearly varies across developing countries and within developing countries. In assessing links with poverty, it is thus necessary to gauge the significance of tourism in countries where poor people are concentrated. Francesco Frangialli, Secretary General of the World Tourism Organisation, indicated in his speech at the World Summit on Sustainable Development[8] that in many of the least developed countries with high levels of poverty, tourism is a significant component of the economy and in some countries the highest GDP earner.[9] Likewise, other studies indicate that tourism is significant in most countries with high levels of poverty and current debate surrounding sustainable and/or responsible forms of tourism should focus more clearly on poverty, in order to enhance the possibilities for poverty reduction (c.f. Bennett, Roe and Ashley's 1999; DFID 1999; Ashley, Roe and Goodwin 2001). While Cuba is not classified as a country with high levels of poverty, it is a poor country which has acknowledged that tourism can benefit its people. The intersection of poverty and tourism in Cuba is currently different from what is taking place with pro-poor tourism throughout countries in Africa and Asia, as it is not part of the mainstream development agency work. Nevertheless, it offers an insight into alternative emerging trajectories of tourism as a development tool, which may well become useful to Cuba if its government undergoes neoliberal economic change in the future, especially since Fidel Castro retired as Cuban President and Democrat Barack Obama was elected as President of the United States.

It has long been argued that because foreign private sector interests often drive tourism, it has limited potential to contribute to the alleviation of poverty in developing countries. Indeed it can disadvantage the poor in many ways. Tourism is renowned for its high incidence of revenue leakage, and generally, the revenue that is retained in a destination country goes towards high or middle income groups, not the poor. Moreover, tourism has proved to be a highly volatile industry prone to the effects of events such as political unrest, exchange rate fluctuations, and natural disasters, making it a highly insecure livelihood for poor families. Tourism to date has explicitly affected the poor in Third World countries, through displacement, the increase of local costs, decreased access to

---

8 Johannesburg, South Africa, 29 August 2002.

9 Directly and indirectly, tourism drives 5-10 per cent of national GDP, jobs, investment and trade for most countries and even more for high intensity destinations (Lipman 2004).

resources, and social and cultural disruption. However, Roe and Urquhart (2001) suggest that many of the stated disadvantages of tourism can be recognised within many types of economic development in a globalising world. A countervailing view, now recognised by some in the development industry, is that tourism has better prospects for promoting pro-poor growth than many other sectors precisely because pro-poor initiatives actively seek to direct profits to the poor.

The tourism pro-poor paradigm is considered to have unique benefits as a developmental tool if managed sustainably, such as its diversity and, hence its ability to build upon a wide resource base and develop in marginal areas with few other options. The tourist consumes tourism at the point of production thereby creating opportunities for local people to sell their goods and services to tourists and become part of the exporter chain. Typically, tourism is highly reliant on cultural and natural capital (for example wildlife), which tend to be assets over which poor people are gaining more control. Tourism offers small-scale labour intensive opportunities compared with other non-agricultural sectors. There are a higher proportion of women employed in the tourism sector compared with other sectors. Given the fact that tourism has steadily increased in poor countries these factors indicate that tourism can indeed benefit the poor, Cuba is no exception. Clearly styles of government will have an important impact but some of the general issues described thus far are relevant to capitalist and socialist countries alike in the new global order. NGO study tours and tourism in general play a huge role in Cuba's development.

One clear understanding from pro-poor tourism case studies thus far, is that in order for it to succeed, the entire tourism destination needs to be successfully developed with appropriate infrastructure, for example, transportation into the area as well as airports. Three fundamental activities guide pro-poor tourism: increasing access by the poor to economic benefits of tourism by expanding the business and employment opportunities specifically to the poor; secondly, addressing the negative social and environmental impacts associated with tourism such as a loss of access to land and resources, social disruption and exploitation; finally, policy reform through the creation of a planning framework that eliminates some of the barriers to the poor through the promotion of their participation in the planning stages of tourism initiatives and by facilitating partnerships between the private sector and poor people. Poor people must be included in the decision-making processes of tourism. A holistic livelihoods approach needs to be incorporated recognising the economic, social, environmental, short and long term concerns of the poor rather than simply focusing on income and jobs. Pro-poor tourism needs to draw on lessons from poverty analysis, environmental management, good governance, small enterprise development and rights-based development.

Pro-poor growth strategies are linked to rights-based approaches to development in that their focus is on poor people and their empowerment. It is rights-based development that I consider in detail, for exploring the tourism-development nexus, and its application in the streets of Havana today.

The Merging of Human Rights with Development Discourse and Practice

> Human rights give a language of political contract to matters of poverty, injustice, and armed violence. Rights talk stops people being perceived as 'needy', as 'victims', and as 'beneficiaries'. Instead it enables these same people to know and present themselves as rightful and dignified people who can make just demands of power and spell out the duties of power in terms of moral and political goods. (Slim 2002: 3)

Rights-based approaches to development provide a framework and give weight to what is taking place in Cuba with NGO study tours. Rights talk gives development a moral vision (Slim 2002: 3). It is pious and makes people feel virtuous. It is this framework along with theories of globalisation and tourism, which I shall draw on in the later sections of this book to make an argument for ways in which tourism can be utilised by development. In this respect tourism moves beyond simply boosting the economy to having wider implications for local people.

The human rights paradigm is yet another shift in development theory and practice. In the last decade, rights-based approaches have gained more attention in development discourse. There appears to be some agreement upon basic elements, but there is no single, universally agreed rights-based approach (Ljumgman 2005). Rather, "there are plural rights-based approaches, with different starting points and rather different implications for development practice" (Cornwall and Nyamu-Musembi 2004: 1415). Essentially human rights are cited as both the moral and legal entitlements to basic well-being and dignity (Ljumgman 2005). They are the "social and political guarantees necessary to protect individuals from the standard threats to human dignity posed by the modern state and modern markets" (Donnelly 1989). This brings us to consider dignity as an inalienable aspect of human existence and raises interesting questions for how development either improves or diminishes human dignity. The human rights framework is unique in its ability to intersect with various aspects of development (Van Tuijl 2000). As such, it provides a paradigmatic foundation for global communication at both the human and institutional levels, because as Van Tuijl (2000: 619) argues, "people need structures and language to relate to each other, to frame what they want to identify as progress or development in their lives".

Recognition of the importance of human rights and the necessity to incorporate human rights into development discourses and practices emerged as a moral imperative to combine economic, social, cultural, civil and political rights. In 1948 the *Universal Declaration of Human Rights* (UDHR) identified two sets of rights: (1) civil and political rights; and (2) economic, social and cultural rights, but by 1950, the drafting of a single convention raised problems in reconciling the two sets of rights. Because the nature of the obligations was very different, two covenants were adopted in 1952: the *International Covenant on Civil and Political Rights* and the *International Covenant of Economic, Social and Cultural Rights*.

This separation influenced the institutional development of the UN. In 1986 there came the *Declaration on the Right to Development,* which brought back together the two sets of rights by effectively stating that they are indivisible. This was a turning point within development discourse and practice, because it acknowledged that we cannot have economic and social development in a situation where civil and political rights are not guaranteed, and similarly, we cannot have civil and political rights where poverty exists.

The distinction between the two categories of rights developed into a fierce ideological debate between Socialist States and the West, with human rights becoming a major battleground in the Cold War (Hamm 2001: 1006). What is now termed a rights-based approach to development emerged within development discourse during the post-Cold War period in the early 1990s (the Copenhagen Summit on Social Development in 1995 exemplifies this). It infers that prior to World War II, development and human rights were seen as separate domains and that this approach represents the convergence between development and human rights. With a rights approach "the boundaries between human rights and development have disappeared, and both become conceptually and operationally inseparable parts of the same process of social change. Development comes to be redefined in terms that include human rights as a constitutive part" (Uvin 2002: 6).

The UN has continued reinforcing the principle that democracy, human rights, sustainability, and social development are interdependent. The convergence of rights and development gained further momentum with UNDP's *Human Development Report* (2005), which offered a persuasive argument for an integrated approach by employing the principles of international human rights to advance dignity and well-being (Ljumgman 2005). The UNDP elucidates the conceptual basis for a rights-based approach stating "the central goal of development has and, will be, the promotion of human well-being. Given that human rights define and defend well-being, this approach to development provides both the conceptual and practical framework for the realisation of human rights through the development process" (Cornwall and Nyamu-Musembi 2004: 1426).

The value of incorporating rights into development practices lies broadly in its normative and ethical aspects. A human rights approach to development sets out a vision of what *should* be, thereby providing a normative framework and it makes an ethical and moral dimension imperative to development assistance. Hamm (2001) recognises that a rights-based approach does not necessarily guarantee success but it does bring about important changes for the ongoing sustainability of a relationship between development and human rights in this era of neoliberal globalisation and 'empire'. It is not my intention to engage longer philosophical discussions on the nature of morality but it is my intention to recognise the implicit moral dimension required in such approaches. They become even more relevant when we shift our focus to tourists in Cuba.

Many would concede that this approach is not new.[10] Examples of elements central to a rights-based approach, that have been taking place for decades, include development work that engages local people in a more active process of social transformation; and advocacy and empowerment projects towards capacity-building. Demands for participation by disadvantaged social groups in decision-making in order to promote bottom-up and people-centred development have been at the core of development literature since the late 1970s. Often development organisations, international financial institutions, and sometimes even NGOs have a somewhat formal notion of participation, which can mean informing local people of development projects rather than including them at inception (Gardner and Lewis 1996). But the shift in focus, where internationally agreed legislation supports these practices, *does* "change the way in which they come to be viewed by development agencies and national governments" (Cornwall and Nyamu-Musembi 2004: 1418), because a rights approach implies that participation is a right and, thus, includes control of planning, process, outcome, and evaluation of the path of development (Hamm 2001: 1018-1019).

The tour groups to Cuba for a number of reasons, at both agency and State level, are heavily participatory giving tourists a close sense of precisely what it means to see this level of engagement in development as a 'right'. The reframing of participation as rights is about assisting people to claim their rights *and* strengthening the capacity of those responsible to fulfil their duties thereby it shifts the development framework from assessing only the needs of beneficiaries. The Cuban organisations that take part in the tours can be considered to also be strengthening their ability to 'do' development in Cuba.

Some critics assert that a rights-based approach assumes that all human rights are universal. They question whether economic, social, and cultural rights are truly reconciled with civil and political rights. This potentially leads to a dilemma of competing claims to accountability and responsibility. One of the main criticisms of rights-based approaches is that this convergence of human rights with development caters to Western interests and in effect is yet another form of top-down Western derived development. In the context of tourism development some argue that far from expressing universal concerns, rights discourses are inherently biased; "Western proponents of human rights often seem oblivious to the manner in which their discourse reflects their own cultural origins in modern capitalist societies" (Smith and Duffy 2003: 81). It could thus be argued that Western institutions and interests manipulate rights discourse to suit their own economic and political

---

10   Organisations such as the ILO, UNICEF, Save the Children, UNDP, OXFAM and Care have adopted and trialled aspects of a rights-based approach. Indeed their experiences have contributed to the evolution of human rights approaches to development with perhaps the most extensive advancement of international human rights being their integration into all UN activities – thus making the conceptual link between human rights, democracy and development. In the last decade bilateral donor agencies such as Sida and DFID have also begun promoting rights perspectives within their agendas.

agendas. Smith and Duffy (2003: 77) contend that "contemporary governments ... have ... deployed discourses of rights in order to claim the moral high ground – that is, to argue that their actions are guided by ethical principles that transcend self-interested political considerations. However, their (political) actions often speak louder than their (moral) words". From a post-development perspective, Escobar argues that the incorporation of human rights into development could be considered another example of Western colonialist interventionism imposing a paradigm of moral ideals on the developing Third World. "The attempt to globalise a doctrine of human rights carries within it an implicit belief in the West's own moral superiority and actually serves to marginalise alternative moral traditions in the Third World" (Smith and Duffy 2003: 84).

Another criticism of rights approaches argues that much discussion of incorporating human rights into development policy and practice amounts to little more than rhetorical repackaging (Uvin 2002: 5). Slim (2002) implies that much of this scepticism arises from a feeling that only the development language has changed and simply 'talking rights' is nothing more than an improved discourse of little use to those living in poverty around the world (Slim 2002: 3). This doubt is exacerbated by the World Bank which positions itself as a proponent of the rights-based approach. The Bank argues that through its work it has always striven to create an environment in which human rights can flourish. But while the Bank embraces the language of rights, it does not change its operations or economic policy and, indeed, we continue to see the World Bank push for the privatisation of essential services, such as water, despite over two decades of experience indicating that such policies further impoverish poor households. Dignity becomes a highly marginalised commodity in such situations. Uvin (2002: 4) warns us that when we read in countless documents of the need for participation by poor people in the processes that affect their lives, it is usually simply a discursive appropriation that amounts to misrepresentation in order to benefit from the moral authority of human rights discourse. He argues that the development enterprise strives for this moral high ground so as to mobilise resources. Others argue that rights talk reflects the moral superiority of the donor with regard to what would be in the best interests of the South (Cornwall and Nyamu-Musembi 2004: 1420).

However, criticisms such as these should not detract from the importance of rights. Rather they highlight some of the difficulties inherent in trying to uphold universal human rights. Some NGOs are using human rights talk in a visionary way to give the subaltern voice power, and in some instances they are challenging Western power very effectively (Slim 2002). However, the moral imperative of rights-based approaches should not be dismissed as purely rhetorical or 'improved discourse'. It can function very differently when spoken by different people:

> The same language of rights that may be rhetorical fluff in one place may be words of extreme courage and radical change in another. The power of speech is the power to name and define things. Rights talk in Washington or Paris might be used piously as new words for the same old liturgy in the cathedrals

of international trade and development ... It represents the power of re-dressing
rather than power of redress. But from another place (a slum or the scene of a
rigged election) and spoken from another voice (that of a poor man or a woman
land rights lawyer) the same words of rights talk could function prophetically as
a demand for redress to change and challenge power. (Slim 2002: 3)

For Slim (2002) the shifts in discourse which have positioned development as
charity, modernisation, and now human rights, is enormously significant precisely
because it has the ability to politicise development.

The critique that rights-based development amounts to little more than
discourse – as evidenced by the World Bank's lack of institutional and policy
change – is what gives rise to the emergence of a global countermovement. The
failure of many national governments and international institutions to ensure
human rights, leads to the perennial question concerning rights-based approaches
to development: who is responsible for providing 'rights' – community, NGOs,
national governments, or international bodies? Indeed even though rights-based
discourse is employed within international aid, far too often donor countries do
not see themselves as bearing any responsibility in the realisation of these rights
(Cornwall and Nyamu-Musembi 2004: 1423-1425). Thus we have seen a rise in
new social movements, who represent the subaltern voice and demand human
rights.

It is clear that to think about alternatives to development requires redefining
theoretical and programmatic notions of development. This possibility relies largely
on the actions of social movements and can thus best be achieved by building upon
their practices. It is essential to link proposals of alternative visions of society with
the ongoing work of social movements (Escobar 2005: 344). Discourses pertaining
to social movements identify two orders – the old and the new. Continuities
exist between both at the level of theories of politics, development, and that of
popular practices (Escobar 2005: 344-345). The 'old' typically refers to analyses
of modernisation or dependency; to politics centred on struggles for control of the
State. The 'new', by contrast, focuses on social actors. Rather than a search for
grand structural transformations, new social movements strive for the construction
of identities and greater autonomy through modifications in everyday practices
and beliefs (Escobar 2005: 344).

Conceivably social movements as symbols of resistance to the prevailing politics
of knowledge and power, provide avenues for the re-imagining of the Third World.
A mosaic of new social movements supported by NGOs to actively mobilise around
issues of human rights, resist and contest the orthodoxy of economic globalisation.
This is because many consider that the failure of development combined with
the nature of current global neoliberal tendencies undermines human well-being
and dignity. The NGO campaign for a rights-based approach to development that
spearheaded the World Social Development Summit in Copenhagen supports this.
Indeed a rights perspective is helping NGOs to respond to some of the challenges
they face to become more relevant in the development arena (Van Tuijl 2000: 617).

It is largely NGOs that take the initiative in guiding the grassroots development activities (McMichael 2004).

New social movements have responded to the failure of development and the social exclusion exacerbated by globalisation through attempts to reframe development as a question of rights. Human sustainability requires conservation of community in inclusive terms rather than the exclusive terms of economic globalisation (McMichael 2004: 281, 307). Many of the people negatively impacted by the processes of development and globalisation look to NGOs to represent their needs. Thus many NGOs support new social movements, whose focus is to defend the rights of the oppressed or to create local and sectoral sites of resistance – ecological, feminist, ethnic, human rights. Together with union organisations and anti-WTO groups, new social movements converged in Seattle in 1999 (Sader 2004: 259) to speak out against the disempowering dynamics of the globalisation project (McMichael 2004: 281). Other forms of resistance to the globalisation project include growing consumer advocacy, for example, against sweatshops and child labour. As I demonstrate, part of that growing consumer advocacy includes new types of tourism selected for their 'educational', 'responsible' and 'political' overtones and tourists' 'moral' sense of doing something 'good'. What we see happening with new social movements is a political interconnectedness, globally and nationally, that at its core has a human rights focus on development. This wave of concern can also be discerned in the rise of the types of tourism I focus on here and in the types of networks they propagate.

Despite criticism of rights-based approaches to development, we need to consider human rights because processes of neoliberal economic globalisation threaten the social standards of those in developing countries who are powerless. Human rights can offer protection from negative outcomes of globalisation (Hamm 2001). It offers a cogent agenda with which to consider tourism as a means to sustainable development. A rights-based approach to tourism provides both the conceptual and practical framework for the realisation of empowering the subaltern voice. NGOs such as Oxfam and Global Exchange assisting local people to get involved in tourism have the capacity to support people to claim their right to participation and thereby strengthen their sense of human well-being and dignity. This aspect of the NGO tours that are the focus of this book gives them a moral underpinning, because not only are they seen as educational, but also as rights-based and thus sustainable. In this sense, we might re-frame our typologies and imagine the emergence of 'rights-based' tourism. Global conjunctions, historical perspectives of development, and moral imperatives of recent development trends inform tourism and what is taking place in Cuba. The next chapter draws on notions of sustainability and participation to consider issues of morality and ethics underpinning new directions in tourism.

Chapter 2
Moral Routes to a New Tourism

Over the last few of decades, there has been an increase in forms of tourism which claim to be 'alternative', 'responsible' and 'sustainable'. This 'new tourism' represents a departure from 'mass tourism'. The tours to Cuba are a clear example of the ongoing evolution of styles of tourism and their instrumental effects. 'New tourism' draws on tropes of sustainability in which a concern for development – environmental, cultural, social and economic impacts – and the participation of local people in tourism decisions are central to its tenets. Viewed in these terms, NGO study tours define a rights-based tourism. Sustainable, responsible and appropriate forms of development are dominant themes implicit in new forms of tourism. They are positioned by the press and tourism operators as 'morally superior' and 'responsible' alternatives to mass tourism, appealing to the growing middle classes. A sense of superiority over traditional mass tourism comes about because such tourists consider that they embody personal qualities such as cosmopolitanism, experiential education and ethical concerns about environment and human rights and so forth. Some theorists (c.f. Cohen 1987; Urry 1990) contend that this trend indicates a consumer reaction against mass tourism and that the emergence of these specialised markets forms part of the post-Fordist mode of consumption. The contemporary cultural processes which underpin the rise of the new middle classes in the West directly promote 'new tourism' in developing countries. This in turn moves analysis beyond tourist motivation and satisfaction in order to show how sustainability, 'new tourism', globalisation and development intersect.

A convergence between development and tourism is clearly evident in the increasing involvement of development agencies in tourism. The notion of sustainable development within tourism in turn leads us to consider issues of morality and ethics. To date, debates about development and tourism synergies do not consider notions of the moral. It is not my intention to engage a philosophical discussion on the parameters of morality. Instead I will discuss NGO study tours as examples of 'new moral tourism' and as the means by which development and tourism are finding mutual interest in their conjunction. Furthermore, a moral theoretical framework is useful for investigating this particular new tourism niche in Cuba, as it sheds light on the instrumental outcomes of tourism as a developmental tool. This is the paramount question underpinning this convergence as we face a changing world where almost everything has become more global and this "world-in-motion" produces disjunctures, disconnections and exclusions (Appadurai 1996; 2001).

New forms of tourism and the niche markets they characterise raise questions of sustainability that have become a corollary element of modernisation. The claim of many of these 'new' tourisms to be sustainable and appropriate alternatives to mass tourism is in response to the perceived 'unsustainable' nature of much tourism development to date and a general move globally towards environmental concerns. Although forms of 'new tourism' are relatively minor when compared with all forms of holiday travel to developing countries,[1] their importance should not be disregarded – they are highly significant in terms of their presence as indications of contemporary change both in their mode of emergence and in their style of operation. As the data from Cuba will show us, this style of tourism forecasts an increasing connection between globalised social movements and consumer engagement in social change and endogenous development endeavours.

New Tourism: Values and Characteristics

> Society Expeditions is a travel firm based in Seattle, Washington, US They run voyages to remote areas, ostensibly for people who are interested in anthropology, ecology and unspoilt environments. I joined them for a month in late 1990 as a specialist guide and lecturer in natural history. We embarked in Bali and sailed to remote destinations through eastern Indonesia, including Komodo Island, then to Broome, across Northern Australia, to the Aru Islands, Irian Jaya and Papua New Guinea. A lecturer on board had lived with the Asmat of Irian Jaya for five years. He was a great favourite with the passengers and involved them in his talks by getting them to join in Asmat chants. The Asmat have particular customs … e.g. they adopt visitors by offering their nipples to suck and they used to, if not still, take heads. … I was concerned … that the Asmat were being presented as sexual and anthropological curiosities. However most passengers didn't seem to view them this way; they were mostly concerned about behaving in a manner acceptable to the Asmat. Meaningful interaction was encouraged in many ways. Staff discouraged handouts to individual villagers instead educational and medical supplies are handed to the relevant person such as teachers or headman. (Denise Goodfellow – Society Expeditions, from web promotional literature)

> The name of the village is Dang jia shan, or the mountain of the Dangs. It is located in the arid area in the northern part of the Loess Plateau. The cave dwellings are built on the cliffs according to the natural topography of the mountains. We stayed in the village with a host family in their cave dwellings. The water supply in the village is lacking, so although water for drinking and washing your hands and face is available, there are no showering or bathing facilities. A bath house is located in a town 5 km from the village. We walked there twice in our two

---

1 Ten to 12 per cent of all tourism is attributed to new tourism (WTO 1995b: 28).

weeks there. I was on an Earthwatch expedition to study a small old village of thirty households in northern Shaanxi province of China. The village is at a critical point, because a State-sponsored reforestation program will require the villagers to be relocated to the river valley near the highway. The reforestation program has been carried out for a year already. Funded by the government, the houses of the new settlement have been built, although the villagers have not yet made their move. This expedition was to assist in preserving the traditional built environment as well as the villagers' way of life. In the project, we made a full-spectrum record of the village before it is left behind. So we recorded data on geographic, ecological, architectural, and folkloric information. We put in long eight hour days of field research and did not break for lunch until three o'clock. Each evening we would have a team meeting for an hour then free time for another hour and supper at 9pm. We had five teams and I was part of the folklore team documenting the cultural environment. I observed and took notes to record the daily life in the village, held interviews with the villagers through interpreters, video taped activities of art and craft production, and took digital pictures. It was just an incredible experience to live in the village and share in their way of life and to be a part of documenting their culture before they are displaced by 'modernisation'. I got to experience something that is inaccessible to tourists. It was really hard work but it was also the holiday of my lifetime. (Ruby – Earthwatch Expeditions, from web promotional literature)

Maybe I'm a frustrated cultural anthropologist, Rochelle, I'm not sure. But I like to learn about other cultures and to study them and understand them and not to try to inflict my values on them. That's why I was interested in the Elderhostel travel programs and Global Exchange's Reality Tours. I've been on a number of expeditions with Elderhostel because a significant part of their mission is to build bridges through learning, travel and cultural exchange. For instance, the trip I took to Costa Rica was an ecological sensitive program and we stayed in eco-tourist lodges, which were in the woods. We met with Costa Rican scholars to discuss what grows there and what's natural and what the birds are. We met with local people to gain a better understanding of their experiences and ah they talked to us about how they make a living and that kind of thing. (Grace – Women's Delegation to Cuba, personal communication)

The above anecdotes are examples of a 'new' tourism and its new forms of community engagement and experience. While international tourism has always been dynamic, adapting to historical and cultural milieu, a review of academic tourism literature and brochures reveals an increasing number of labels, such as 'alternative' and 'new', for travel experiences or products that explicitly position themselves in opposition to mass tourism. While there is no universally accepted definition of what constitutes 'alternative tourism', Cohen (1987: 13) suggests it has evolved from two distinctly Western ideological preoccupations; the first being the rejection of mass consumerism by the counter-cultural movement. In the context of tourism, these people engaged in a quest for more authentic experiences by avoiding the amenities of the tourism establishment and attempting to travel independently of the existing tourism infrastructure (Cohen 1987: 14). The second ideological preoccupation underpinning the development of alternative tourism is the neo-Marxist critique arguing that international trade, development efforts and even foreign aid strengthened a neocolonial dependency by the Third World on the industrialised world, thereby perpetuating their lack of development (Cohen 1987). Hinch and Butler (1996) unequivocally state that the contact between the developed and developing worlds has traditionally been characterised by the exploitation of indigenous people for the benefit of non-indigenous groups. The outcome is that indigenous people are involved in a constant struggle for their own cultural survival. Mass tourism contributes to the appropriation of the amenities of developing countries and the exploitation of people and cultures for the recreation of Western tourists. In combination, the rejection of mass marketing and critique of new-imperialism is driving alternative tourism and the increasing number of concerned Westerners who now seek social, cultural, and educational forms of tourism. The question as to whether new forms of tourism bring about the same problems as those associated with mainstream tourism is not yet answered (Munt 1994b) but Butcher (2003) argues that new forms of tourism are not fundamentally different and suggests they do indeed bring about the same problems.

Alternative or new forms of tourism are primarily associated with values designed to reduce negative impacts and enhance sustainability, and are characterised by a desire for 'authentic' experiences involving local people in decision-making (Lash and Urry 1994; Brown 1998; Weaver 1991; Prosser 1994; Munt 1994b). 'New tourism' is also individual, flexible and segmented in nature (Poon 1989), providing opportunities for a level of engagement in fields such as archaeology, development, anthropology, ecology, conservation and science (Weiler 1991). The incorporation of an educational component encourages some travellers to favour those trips that cater to their intellectual aims to seek opportunities for personal growth and development. Educational travel warrants further attention in order to identify the potential benefits to host communities, participants, and the tourism industry (Weiler 1991). These benefits directly encourage hundreds of people to go to Cuba each year with the tour subsidiaries of Oxfam and Global Exchange.

New forms of alternative tourism claim to be more sustainable and individual in nature. These flexible forms are in part driven by demand, but also by need. Environmental concerns, new global 'best practice' of flexible production, and most recently, a move by development agencies to consider tourism as a development tool are emerging as the most significant externally-driven changes to tourism, and new middle class consumers considered to be the most significant internal architects of change within tourism (Poon 1989).

Transformations in consumer behaviour, values, and motivations provide the fundamental guiding force for new tourism (Poon 1993). New consumers acquire changed values and lifestyles which are oriented towards the environment and the ethical consumption of tourism. They have motivations which are related more to active, rather than passive, vacation activities, the search for escape, and for the authentic (Munt 1994b). Some authors consider that tourists are increasingly using vacations as an expression of identity through holiday styles and destination choices (Lash and Urry 1994; Brown 1998; Weaver 1991; Crompton 1993). The changing lifestyles of these people are creating demand for more targeted holidays which cater to their specific situations, while changing values are also generating demand for a more environmentally conscious and nature-oriented holiday. 'New tourists', according to Poon (1993), consider the environment and the culture of the destinations they visit as a key part of the holiday experience. Expanding on this idea, Butcher (2003) labels these tourists 'new moral tourists', because they see themselves as morally superior compared with other tourists, owing to the values and characteristics associated with 'new tourism'.

In the 1990s, questions were being asked about exactly who these bearers of new moral values were and how their lifestyles were reflected in travel (Munt 1994b: 50). Some answers are becoming clearer. Links have been made between postmodernism and a growth in what has been labelled the 'new middle classes'. There has been a swift expansion of social plural classes in the West. Bourdieu (1984) referred to these new middle classes as 'new cultural intermediaries', identifying them as agents of the cultural change inherent in postmodernity. The growing middle classes have been identified as key social groups in initiating and perpetuating new cultural processes and consumption patterns. Travel choices are a key factor in these processes, as these social classes are both important consumers of holidays in developing countries and key groups in promoting notions of sustainability – a fundamental element underlying forms of 'new tourism' in developing countries. Mowforth and Munt (1998) point out, though, that this does not indicate that tourism in developing countries is purely the product of the cultural consumption choices of the new middle classes in the West but simply that the latter have a significant impact.

Similarly, Crompton (1993) draws parallels between the rise of the new middle classes and the growth in consumer capitalism and its emphasis on 'lifestyle'; consumer preferences "make up the lifestylization of consumption" (McCabe and Stokoe 2004: 604). This can be seen in the shift to increasing consumption of services. In this context, lifestyles include places where people choose to live, the activities in which people engage and, clearly, the holiday choices people make (Crompton 1993). In a so-called 'postmodern'[2] era of cultural change, we all attempt to convey our identity to others through the items and services we consume within our lifestyle choices. Thus tourism styles are fundamentally linked to constructions of modern middle-class social identities (Munt 1994b) and a potentially 'good' or 'moral' self (Matless 1995). Consumers have the power to effect change and force companies to engage in more ethical behaviour by exercising choice in purchasing products and services that are considered more ethical. Importantly, consumption practices play a part in our constructions

---

2   See Thurot and Thurot 1983; Urry 1990.

of identity and our sense of a moral self (Matless 1995). For Urry (1994: 235), "identity is formed through consumption and play. It is argued that people's social identities are increasingly formed not through work, whether in the factory or at home, but through their patterns of consumption of goods, services and signs". Where previously in industrialised societies identities were formed relative to the workplace, today, through consumption, we can tailor our individual identity through our tastes and moral values. In this way, the consumption of tourism has itself gained a moral dimension.

Through descriptions of travel experiences, tourists create versions of the social world intending to demonstrate a profound form of cultural competence that is central to the moral achievement of identity (Baker 1997). Importantly, a growing number of people are involved in the emerging sociopolitical movements that reflect lifestyle and identity. These movements focus on a range of issues such as the environment, human rights, religion, alternative lifestyles, women's rights and so on. Habermas (1981) refers to the emergence of these sociopolitical movements as 'new politics'. Thus we see that growth diversification incorporates sustainability. Following the embrace of sustainability by development practitioners and theorists, tourism is following suit with the emergence of ecotourism and more recently 'new tourism'.

NGO study tours are one way in which these new middle classes may actively engage with social/political/environmental movements in order to understand and experience issues in developing countries. As I will describe, the people participating in development-oriented tours – many of which I guided through Cuba – demonstrate a high involvement in social movement participation. Their level of support in such movements and of charities, development agencies and conservation groups indicates their commitment to and, involvement with, development issues. It also reinforces how meaningful the experience of participating in a NGO study tour is related to other aspects of their lives, such as charity giving, volunteering or activism. A brief snapshot of the support by those tourists travelling on NGO study tours illustrates that it varied from financial contributions, paid membership, volunteerism, activism, and previously participated in NGO study tours. The following quotes show us the tenor of their engagement:

> The charities I support actively with fundraising, lobbying, organising and public speaking include Amnesty International – where I also convene the Mount Waverley group – and Oxfam. In addition, I'm a paid-up member and regular contributor to Australian Conservation Foundation, Greenpeace, Medecins Sans Frontieres, Fred Hollows Foundation and lots of Australian welfare agencies and some medical charities. (Francesca – Oxfam Community Aid Abroad Tours)

> I participate actively (speaking, writing, rallying etc.) in some groups involved with healthcare reform and with reform of the criminal justice system. I frequently write letters for Amnesty International. I give financial support to quite a few charities, ranging from Oxfam and Save the Children and Amnesty

International to museums and a wide range of activist organisations. (Henrietta – Global Exchange Reality Tours)

I support the women's movement, anti-war movement, different labour issues, Central America solidarity, um I ah go to benefits, like I went to a Cuban Peña which was to raise money for aid to Cuba ... my husband is also an activist. Yeah so over the years we've gone to a lot of benefits. For a long time that was our primary social life; doing political work and going to charitable benefits ... I sing in a women's chorus and we do outreach to groups like Habitat for Humanity, Martin Luther King Day, International Women's Day, an AIDS hospice and a nursing home. (Stella – Global Exchange Reality Tours)

External forces beyond identity politics are also relevant to the emergence of new forms of tourism. Post-Fordism (Lash and Urry 1994; Mowforth and Munt 1998; 2003) is one explanatory framework; it refers to the move from mass production and consumption to considerably more flexible systems of production and organisation. Most importantly, post-Fordism indicates changes in the way society consumes goods and services and the rapidly changing nature of tastes and trends which lead niche markets. These ideas can be readily applied to 'new tourism'. Indeed, Poon's (1993) thesis of 'new tourism' argues that the development of the tourism industry has closely followed that of the manufacturing sector by initially adopting the same principles of mass production and, more recently, moving towards flexible production as new consumers assert their lifestyle choices. Hence, there has been an obvious shift away from the Fordist regime of mass tourism and packaged holidays and a steady move towards specialised markets, independent holidays and small group tours to developing countries.

Recent shifts in tourism therefore need to be considered in terms of the global, as this characteristic is what clearly typifies the 'new'. The globalisation literature offers a nuanced insight into the growth and development of tourism in developing countries because, as Tsing (2000) contends, a global framework provides an understanding of the making and remaking of geographical agents and the forms of their agency in relation to movement and interaction. Places are made through their connections with each other. Interrogating globalisation helps us to imagine interconnection, travel and sudden transformation. For Harvey (1989), time-space compression is the driving force behind the economic processes of global change. This view holds that the present movement of capital and information is far more rapid within global networks. To achieve this time-space compression, new markets and products are continually introduced. This process can be seen to occur within international tourism as an ever-increasing number of holiday destinations in Third World countries are developed for tourism. Likewise Third World governments, including Cuba, acquire and actively utilise the growing appetite for a diversified sector.

Thus we witness an increase in flows of capital, information, technology and so forth, which effect the popularity of places as 'destinations'. For example, if a country suddenly descends into civil strife or experiences a natural disaster, people in other countries are informed almost instantaneously. The way events or circumstances in developing countries are represented and perceived within the West can have significant ramifications for the ongoing success of tourism, highlighting its unstable nature as a panacea for development in developing countries, and the ways in which countries are interdependent on one another in this globalised era.

Because most tourists derive from Western countries, discussion of tourism in developing countries needs to address changes in the West and the associated factors that produce new forms of tourism (Mowforth and Munt 1998). Post-Fordism and postmodernism, as theoretical underpinnings, provide important insights in discussions about global change, sustainable development and tourism. They offer a framework within which to understand emerging cultural changes and facilitate an understanding of the relationships between consumption practices in the West and the capitalist orthodoxy. Contemporary Western lifestyles involve an increase in the consumption of services rather than products that are increasingly underpinned by a moral and ethical dimension. This shift to services includes the consumption of tourism services where lifestyle trends are reflected in the selection of holiday destinations. Most importantly, these cultural changes in consumption choices are indicative of uneven power relations in the context of sustainability.

"Tourism is accompanied by constant warnings to be ethical ... [and] is now the terrain of moral codes" (Butcher 2003: 71). The influence of ethics on political discourses is not new and has evolved in different ways that have impacted on tourism practices. The notion of the 'work ethic' (see Marx 1965; Harvey 1973) in particular refers to patterns of work in industrialising societies underpinned by a sense of moral integrity and economic survival. From this grew the notion of a 'leisure ethic' where the pursuit of hedonism became important (Mowforth and Munt 1998). While the work ethic is still stronger than the leisure ethic in Western society, the leisure ethic is gaining prominence due to the spread of services to the wealthy and the rise in paid holiday leave.

More recently, during the 1980s, there was a rise in another ethic – the 'conservation ethic',[3] which can be seen to influence tourism practices and policies through discourses of sustainability, especially throughout developing countries. Some have argued that the new middle classes want the assets and resources in developing countries preserved out of self interest; for their holiday enjoyment. Recently the notion of a new ethic has emerged from the UN World Summit – the 'poverty alleviation ethic', which is the new ethical obligation of an affluent world as evidenced by the Millennium Development Goals[4] and the rights-based approach to development.

---

3   See Thurot and Thurot (1983).

4   The homepage of the United Nations Millennium Development Goals website states "keep the promise" urging 191 UN member states to meet the Millennium Development Goals. The Goals include to eradicate extreme poverty and hunger, achieve universal

These ethical considerations are linked, then, to different forms of tourism and relationships of power. It is important to acknowledge the structures of power in the tourism industry that are a crucial variable in the development and forms of tourism. Mass tourism has often been criticised as a degrading experience for developing countries (Munt 1994b). Butcher (2003) suggests that the same problems are indeed relevant to new forms of tourism which are, more often than not, viewed uncritically. The argument that alternative forms of tourism have developed to address problems within mainstream tourism is often cited as justification for alternative tourisms. However, Fernandes (1994: 4) argues that new types of tourism have arisen as a response by the tourism industry to validate itself, for example, the sustainable and rational use of the environment for Western tourist interests. Hence sustainable tourism as a concept and practice needs careful examination.

Sustainability Discourse, New Social Movements and Tourism

> In a world subject to such enormous ecological and human contingencies, sustainable development, referred to here as sustainability, has evolved over three decades from an environmental issue to a sociopolitical movement for beneficial social and economic change ... Its basic principles are to improve and sustain human well-being indefinitely without impairing the life support systems on which it depends. (Farrell and Twining-Ward 2004: 2)

Within tourism, sustainability has emerged as the dominant discourse driven by moral and ethical underpinnings. A significant number of scholarly contributions on sustainable tourism development have been made over the last decade (Bramwell, Henry, Jackson and van der Straaten 1996; Butler 1999; Hunter 1997; Mowforth and Munt 1998; Stabler and Goodall 1996; Swarbrooke 1999; Wall 1997a; 1997b; Weaver and Lawton 1999). Tourism is broadly defined as sustainable if it is environmentally, culturally, socially or economically sustainable; educational; and locally participatory (Mowforth and Munt 1998). However, as mentioned earlier, the reality is that sustainability is often defined differently because of the range of contexts and power relations within which it is characterised. The most common way in which sustainability is classified is ecologically, through a need to reduce the impact of tourist activities on the environment.

'Social sustainability' indicates the ability of a community to absorb inputs such as extra people (tourists) for short or long periods of time and to continue functioning without the creation of social disharmony as a result, or by adapting its functions so that disharmony is mitigated. 'Cultural sustainability' refers to

---

primary education, promote gender equality, reduce child mortality, improve maternal health, combat HIV/AIDS and other diseases, ensure environmental sustainability and develop a global partnership to development.

the ability of societies to maintain social harmony when lifestyles, traditions and cultures are subject to change through the introduction of visitors with different habits, customs, and modes of exchange. Culture is recognised to be dynamic and thus cultural sustainability refers to the ability of a people to retain or adapt elements of their culture. 'Economic sustainability' refers to economic gain and is usually prioritised over environmental and sociocultural sustainability in tourism development. Importantly, the question of who gains financially and who loses often brings the issue of power control into sharp focus compared to other elements of sustainability. In other words, for tourism to be sustainable it must be low impact and not harm the environment or the local people.

An educational element has often been cited as a crucial feature that distinguishes new forms of tourism from mass tourism. Thus a goal of learning about and understanding the natural and/or cultural environment while on tour facilitates sustainability through local participation with the tours and can be seen to be central to new forms of tourism. Learning about the development issues of the country visited is central to the NGO study tours that are the subject of this book. Some scholars (for example, Bruner 2005) argue that while this may be stated as a goal, in many instances it may not have the desired outcome of positive attitudinal or behavioural change because the tours are too short to allow for any critical engagement. I will demonstrate that the NGO study tours that are the focus of this book offer an intensive program of meeting local people and visiting community projects and result in productive outcomes – despite their brevity. Local participation is another aspect of sustainability that has gained prominence, but the degree of local inclusion in tourism decision-making processes is varied.

'New tourism' represents a marked shift away from mass tourism towards niche environmentally-friendly tourism. New niche forms have emerged to cater for the discerning new tourist. It has been suggested that this reflects two distinct aspects of sustainability: to protect the environment and alternatively to preserve cultural character for the specific benefit of predominantly Western new middle class tourists (Munt 1994b; Mowforth and Munt 1998). The growing body of NGOs with a focus on the environment has directed the emerging global concern for sustainability thereby increasing examination of the role and impact of tourism in relation to 'new tourism'.

I have been suggesting that the growth of 'new tourism' has been traced to recent postmodern cultural changes such as the growth of the socio-environmental movement (c.f. Mowforth and Munt 1998). NGOs have been responsible for transnational campaigning and lobbying on a broad range of issues including anti racism, human rights, conservation, peace, anti nuclear and so on. There is also a rise in the number of organisations dedicated to single causes whether environmental, social or cultural in nature; their focus also includes tourism. The political influence and power that such NGOs can wield may act as a counterweight to the geopolitical forces of transnational companies, supranational institutions and Western government interests.

The growth of socio-environmental organisations is largely driven by the educated, intellectual and socially aware middle classes who use these organisations to pursue and further their own class interests, for example, environmental issues. Eckersley (1989) argues that the new middle classes are the "vanguards of post-material values, the harbingers of sustainable lifestyles" (1989: 161). Inglehart (1977: 1981) indicates that the new socio-environmental organisations mobilise themselves around the importance of education combined with their awareness of social and environmental problems and this produces a new middle class ideology. The rise of morality in the internationalisation of human rights discourses has led to socially-aware tourists concerned with human rights issues because of the high profile of genocide or sex tourism.

Power needs to be clearly defined in order to understand how sustainability is manipulated by the varying interests of international development agencies, international financial institutions, the policies of national governments, non-government organisations, tour operators and even tourists. The relationship between sustainable development and tourism in developing countries requires examination to determine whether, and how, development, including tourism as a developmental tool, can be sustainable. This is necessary precisely because mass tourism in developing countries is typically seen as unsustainable in terms of its negative environmental, cultural and economic impacts. Accordingly, 'new tourism' strives to be environmentally, culturally and economically different from the negative outcomes of mass tourism (Mowforth and Munt 1998: 11), but this needs to be confirmed rather than assumed.

Power and its ability to dominate and damage can be analysed not only by understanding the role of socio-environmental organisations in tourism development but also by understanding the role of governments and the integration of policies of supranational institutions. Developing countries have been subject to the lending policies of supranational institutions, foreign policies of Western governments, and the choices of Western tourists. It is often international bodies such as World Bank, Asia Development Bank, Inter-America Development Bank, International Monetary Fund, agencies of the United Nations, and the World Tourism Organisation which have the power to direct tourism development by money lending and associated neoliberal structural adjustment (Britton 1983: 339). This highlights Western control over much tourism development. But not all developing countries have fallen prey to this form of tourism development. Cuba is a prime example of a small developing country that relies on tourism but is responsive to the issues of power in not borrowing from supranational financial institutions and has instituted a dual economy to keep tourism spending separate from the national economy. This is not, in and of itself, unproblematic and the Cuban strategy does not address issues of power for local people inherent in tourism development or the complexity and multiplicity of outcomes – issues to which I will return.

While there is recognition that tourism inevitably takes place in the context of great inequality of power at numerous State, community and individual levels, moralities about damage to culture and environment have become significant within the sustainability discourse. In turn, moral underpinnings are prominent in driving the 'new tourism' agenda.

Tourism and Morality

The discourse of sustainability underpins increasing support for more ethical behaviours in tourism and what Butcher (2003) refers to as a moralisation of tourism. New moral tourism, he argues, has been building as an effective agenda since Agenda 21 was tabled in Rio in 1992. As I have suggested, this moralisation stems from a growing concern about the damage caused by mass tourism to both the environment and cultures. In addition, it may also be driven by educational desires to learn about other cultures. As Poon (1993) reminds us, those people participating in 'new tourism' forms are experienced travellers, educated, independent, flexible and environmentally concerned. Poon establishes that what is central to 'new tourism' is a particular ethical imperative. It is positioned and advocated by NGOs, campaigns and alternative tourism operators as a solution to the magnitude of environmental and developmental problems often caused or exacerbated by mass tourism.

Following this trend, supporters of 'new tourism' maintain that there is a burgeoning market of ethical tourists who seek more individualised, environmentally and culturally sensitive forms of holidays. In many respects this growing market is a response to "key features of their moralised conception of leisure travel ... a search for enlightenment in other places, and a desire to preserve these places in the name of cultural diversity and environmental conservation" (Butcher 2003: 8).

New moral tourism has arisen from a diverse range of organisations and campaigns. Global conservation organisations like World Wide Fund for Nature and Conservation International increasingly view ecotourism as a means of achieving conservation goals. Ecotourism's ethical credentials come from its ability to combine conservation with limited development goals. While mass tourism yields more economically than ecotourism, it is considered less ethical due to its environmental and cultural impacts. Other projects, advocacy campaigns and NGOs promoting an ethically conscious tourism include Proyecto Ambiental Tenerife, Save Goa Campaign, the International Ecotourism Society, Worldwise, Tourism Concern, Campaign for Environmentally Responsible Tourism, Green Globe, Earthwatch, Ecotourism Society, the World Wildlife Fund, the Sierra Club and more. Goals range from supporting tourism that leads to intercultural encounters, facilitating joint learning processes, promoting the ethical credentials of green holidays, conserving fragile ecosystems, supporting endangered species, preserving indigenous cultures, and developing local sustainable economies, encourages prospective tourists to travel with a personal and global purpose, and produces ethical codes of conduct for travelling; to share, learn and grow with the local people.

The discussion so far, highlights the parameters of a dynamic and complex new array of variables massing within a moral framework. New moral tourism implies that the tourist can engage in travel that is meaningful and enlightening and it demands that people regard their holidays differently in order that they impact positively on the environments and cultures of developing countries. While it is clear that there is an increase in new moral tourism, clearly not all tourism has these ethical and moral underpinnings. Other tourism niches profoundly lack a sense of morality, such as sex tourism still common throughout the developing world. For our purposes, the rise in ethical tourism codes and NGO campaigns and activities illustrates a distinct agenda that underlies the creation of a niche market. Advocacy by diverse organisations and new social movements, which creates validity and legitimises participation, is an important aspect of new moral tourism.

Some might argue, however, that the involvement of development agencies in taking tours to their development projects can be seen as an "ultimate aestheticisation of reality, through which racism and class struggle can be enjoyed ... in its new wafer-thin disguise as a more ethical and moral pastime of the new bourgeoisie" (Munt 1994b: 59). Butcher (2003) argues that the assumptions implicit in new moral tourism are rarely challenged. Butcher's *The Moralisation of Tourism: Sun, Sand ... and Saving the World?* (2003) critiques the assumptions that mass tourism has been destructive and that ecotourism is somehow ethical. He argues that both eco- and culturally sensitive tourisms are based on false premises of environmental and cultural fragility.

The question arises as to whether 'new tourists' are in fact more individual or moral in ways that distinguish them from the mass tourist. One could say that new moral tourists display a condescending judgement on how others choose to spend their leisure time and that the fact that an individual chooses the same holiday destination as a large number of other people, does not make the person any less individual (Butcher 2003). But this aspect of Butcher's argument fails to acknowledge that new moral tourism reflects a general shift in Western new middle class lifestyle choices. Likewise, Butcher contends that new moral tourists are presented as being interested in the people and the cultures they encounter on their travels while those partaking in mass tourism are not. He argues that this generalisation about mass tourism cannot be made, as the people within mass tourism are diverse in terms of their interests and motivations. He further critiques new moral tourism on the basis that 'new tourists' are described as educated and thus concerned for the environment and culture of the holiday destination, whereas their counterparts are considered unthinking and unethical. He argues that a lack of engagement in the moral tourism agenda does not indicate that they are necessarily unthinking or uneducated and that it may simply be that they do not consider a holiday as a vehicle for doing good.

I would argue that while new moral tourists are not necessarily more educated, they actively demonstrate a desire to learn about cultural and environmental issues while on holiday. Butcher further suggests that new moral tourism may have an educative aim but this does not necessarily lead to educational outcomes. The aim may be to foster a meaningful experience and stimulate a cognitive response, possibly even positive behavioural change; but this is not guaranteed. He argues that focusing on the culture of the local people at a holiday destination may create a barrier. This barrier is arguably strengthened by its own terms of reference – the need to be more informed on our travels about people and places so that new moral tourism is a stifling etiquette that presents a barrier to discovery. As we will see, the NGO study tours fail to confirm this position. The tours provide forms of social engagement that are productive in a myriad of ways rather than being constraining and limiting.

Butcher's (2003) critique of tourists incorporating a learning component in their leisure time is useful for its analysis of how 'new tourism' places itself as morally and ethically superior to mainstream tourism. However he misses some key elements. His argument rests primarily on how new moral tourism is defined in relation to mass tourism and he seeks to rescue tourism, in whichever form, from being subject to moral codes. While his thesis raises some interesting and valid points regarding the positioning of new moral tourism in opposition to mass tourism, his argument dismisses the significant implications for tourism such as unequal power relations and uneven development which new moral tourism attempts to redress.

The moralisation of tourism attempts to present alternatives to mass tourism that are considered not only better from the perspective of developing countries where tourism is implemented as a developmental tool (typically only economic), but also better for the tourists themselves. For Butcher, consumer preferences about holidays are becoming transformed into moral choices and thereby have significant implications for the tourist destination and subjective experience. As mentioned earlier, although 'new tourism' is a small niche within tourism, it is nevertheless significant in that it is proving to have an influential agenda, encouraging people to 'give something back' in their leisure time and, in many instances, to incorporate an educational component into their leisure time that could have implications for developing countries. New tourism reflects the social change in lifestyle choices of the new middle classes and the agenda of some NGOs to utilise tourism as a means of informing people about issues of concern, such as those pertaining to human rights, and the environment and so on. It is important to point out that these new forms of tourism are born of a moral perspective rather than a product *per se*, although at some point the two merge. New tourism seeks to establish a moral code with which to reform the global tourism industry through better opportunities for the environments and cultures of developing countries and with new moral and ethical codes underpinning tourist behaviour.

The emergence of codes of conduct is a particular aspect of this moralisation of tourism that Butcher critiques on the premise that it wrongly implies that tourists and locals are incapable of negotiating cultural differences. I would argue that, on the contrary and in light of the enormous environmental damage and cultural inequalities stemming from tourism development to date, particularly in developing countries, codes of conduct are a move in the right direction in informing people about environmentally responsible and culturally sensitive ways of travelling. They are one reason for the increase in new social movements towards which people look to guide them in more informed practices and activities in both their day-to-day lives and their leisure time. Butcher argues that codes of conduct are based on the assumption that the uninformed individual is in need of moral guidance by NGOs and new moral tourism operators, which Butcher believes threatens the sovereignty of the individual. It is important to note that codes of conduct and new moral tourism generally, do not represent a rigid set of rules but a set of principles aimed at moving towards better practices within tourism for the local people and environment. The influence wielded by the socio-environmental movement in lobbying for change to the un-environmental practices of major industry has had a degree of success. New moral tourism equally represents pressure on what is now the world's largest industry.

A further relevant issue concerns authenticity, and it might be said that new moral tourists seek leave from modern society through a temporary immersion in a culture they perceive to be more traditional – a search for authenticity (MacCannell 1973; 1976). Arguably, the new moral tourist is more inclined to want to experience the 'real' culture of the destination as opposed to mass tourists who accept the inauthenticity of staged cultural tourist attractions (Boorstin 1969). Urry (1990) claims that the tourist gaze fixes on sites that appear to offer a pre-modern existence from which the tourist can learn a great deal. The advocacy of culturally-sensitive tourism is shaped by the search for the authentic in pre-modern societies. "'Culture' in this sense is often refracted through a distinctly Western lens; one that both elevates the host's culture and at the same time restricts its development" (Butcher 2003: 81). Culture in this context also refers to the profound sense of disillusionment with modern society and its cultural associations of mass consumerism, including mass tourism. "The new moral tourist seeks respite from modernity through a temporary immersion in a culture they perceive to be less sullied by modern society" (Butcher 2003: 78).

This is a highly popularised argument and needs careful consideration to adequately address social change and individual subjectivity well beyond one dimensional and ethnocentric analysis of tourist movements. There are three facets to this perspective according to Butcher. First, change within a culture becomes an attack on culture because, second, culture is idealised as statically rooted in tradition. Third, culture is seen as what makes people different from one another. Culture then can be defined as function, because change to aspects of a culture imposed by tourism is seen as upsetting the functioning of society. Culture can also thus be defined as the past. The association of the present with a lack of authenticity leads to a search for authenticity in the past, which may be found in the cultures of developing countries. And finally culture can be defined as difference. The elevation of cultural *difference* above commonality underpins the advocacy of new moral tourism. Butcher argues that the way the host is viewed through the prism of culture has consequences for development as culture defined in the three forms above arguably becomes a straightjacket for societies that may be seeking economic development. This perspective clearly relates to notions of modernisation, sustainability and tourists' choices of doing no damage and 'giving something back', to prevent or mitigate the effects of change.

Importantly, as Butcher highlights, new moral tourism is not a single moral framework but rather a fluid phenomenon. It represents a tendency to understand tourism in moral terms which are connected to a more acute awareness of environment and culture and a cautious understanding of progress evident in the growth of mass travel. In relation to the search for authenticity, Butcher suggests that new moral tourism is similar to practising amateur anthropology and indeed its practitioners want to be considered as superior to mainstream tourists. As I have been elaborating, this superiority is based on the underlying 'morality' of the new styles of cultural contact. The anthropology of tourism emerged with Valene Smith's concern for cultural contact between hosts and guests. Butcher draws on MacCannell (1976) and Smith (1989) to suggest that, like anthropologists, new moral tourists are interested in learning about the culture of the destination while minimising their own impact on that culture. Both seek to go beyond the spectacle and into the authentic 'backstage' area.[5] The NGO study tours in Cuba, which are the subject of this research appeal precisely to this desire by offering people-to-people contact through an itinerary of seminars and community project visits with local grassroots organisations.

We might argue, that in his or her perambulations the new moral tourist can therefore be seen as the modern day *flaneur*, leaving home to be educated through the very process of travelling. As Benjamin has so famously shown us, the *flaneur* was a stroller, someone who ambled through the Parisian streets of the nineteenth century attuned to the history of the place and in covert search of adventure. Similarly, new moral tourism evokes the 'Grand Tour', which became popular in the eighteenth century among European male bourgeoisie who travelled as an important part of their education and as a reflection of their status in society (Adler 1989). Hannerz (1996: 178) claims cosmopolitanism is "marked by interest in, familiarity with, or knowledge and appreciation of many parts of the world" and "marked by sophistication and *savoir faire* arising from urban life and wide travel". For Hannerz (1996: 103), cosmopolitanism is an orientation and willingness to engage with the Other that necessarily involves an intellectual desire for new divergent cultural experiences. The cosmopolitan demonstrates an eagerness to become involved with the Other, and a concern with achieving competence in cultures that are initially alien; this relates to considerations of self and identity. In this light, cosmopolitans abhor tourists and particularly hate

---

5   For an early critique of the oft-mentioned similarities between anthropologists, tourists and other seekers, see the early critique of MacCannell: Van den Abbeele, G. (1980) 'Sightseers: the Tourist as Theorist', *Diacrities* 10(4): 2-14; Errington, F. and D. Gewertz (1989) 'Tourism and Anthropology in a Post-Modern World', *Oceania* 60: 37-54.

being mistaken for tourists.[6] Cosmopolitans want to immerse themselves in other cultures and they want to be participants; they see themselves as being admitted into local reciprocities (Hannerz 1996). The profile of those people participating in NGO study tours suggests that they are likely to identify as cosmopolitan by nature. Through intense itineraries of seminars and project visits they immerse themselves in the culture and participate in exchanges of knowledge of specific developmental foci. They can return home having achieved a certain level of knowledge and competence in discussing and disseminating information about that culture. They are perhaps early twenty-first century *flaneurs*, wandering the highways of the planet, rather than the boulevards of nineteenth century Paris.

But clearly, as Butcher has argued, this perspective has its limitations. New moral tourism could also be seen as erecting barriers based on a notion of a decentred, self-limiting tourist who is too busy gazing at the Other's culture to empathise with them as individuals, thereby restraining open communication (Butcher 2003). The people whom I observed participating in NGO study tours were, however, engaging with local Cuban people about their development issues. Rather than erecting barriers, these tourists were intent on communicating with the people of the country they were visiting. Butcher wants us to interpret culture as function (the past, as pre-modern) and as difference, and this acts as a straight-jacket for societies seeking development. My concern is that this view is over simplistic, as both tourism and development have the capacity to allow for the fact that cultures will inevitably be made and remade in the context of social change. Conceptions of culture as static are no longer tenable within touristic and development discourses. One goal of the NGO study tours in Cuba is for participants to learn how local people have survived and adapted to a long-standing trade and economic embargo through their development initiatives. This demonstrates that new tourists do not necessarily see culture as static and are in fact willing to learn about the development issues faced by other cultures.

A more penetrating critique is to question the motivations driving 'new tourists', an investigation which the ethnographic work of this research is oriented. Holidays have become important commodities for the new middle classes as a means of proclaiming their worldly status (Munt 1994b). They are ego-tourists participating in relatively expensive styles of travel reflective of, an alternative lifestyle that marks them as different (Munt 1994b: 50), or the ability to demonstrate their taste (Bourdieu 1984). Munt argues that some tourists enact a form of distinction between styles of travel that they convert into class status and that Third World destinations are encountered as places where individual achievement, strength of character and worldliness can be narrated. Arguably, the new middle classes engage in 'new tourism' as a means of accumulating cultural capital to assert a middle class identity (Munt 1994b; Holmes 1998; Desforges 1998; 2000; Urry 1995) rather than for the moral and ethical concerns of sustainable tourism

---

6 For discussion on cosmopolitans and tourists see Errington, F. and Gewertz, D. (1989) 'Tourism and Anthropology in a Post-Modern World', *Oceania* 60: 37-54.

practices. "Travel does not provide cultural capital or other forms of identification in any simple sense, but is given meaning for identity" (Desforges 2000: 11). For the 'new tourists', they can make clear claims for cultural capital of the kind associated with travel (Clifford 1997; Urry 1995). This could be because 'new tourism' is considered by many to embody personal qualities such as a strong sense of morals, cosmopolitanism, education through experience and so forth, supposedly reflecting a moral superiority to those who participate in mainstream tourism. McCabe et al. (2004: 614) advise us that the term 'tourist' is a "culturally constructed category with associated negative category-bound activities and predicates. It is proposed that such formulations work to achieve a sense of moral orderliness in the categorisation of places, activities and people". While I believe that many people in the new middle classes do demonstrate ego-tourist characteristics in their engagement with 'new tourism', their involvement in new social movements indicates a concern for environmental sustainability and the human rights of Third World cultures that is too convincing to dismiss as purely an attempt to simply build cultural capital.

Another useful way to examine new moral tourism is as a form of ethical consumption (Butcher 2003). The idea of packaging capitalist practices as ethical consumption somehow makes them seem benign. However in contemporary political culture it is increasingly evident that people are more conscious of making a difference through their purchases. In a related analysis of the right to access credit, Mohamad Yunus of the Grameen Bank refers to this idea as socially conscious capitalism. However, Naomi Klein's book about resistance through anti-corporate activism in a post-national era warns us that while many initiatives have real merit like whether "your sneakers [are] 'No Sweat'? … Is your moisturizer 'Cruelty-Free'? Your coffee 'Fair Trade'? … the challenges of a global labour market are too vast to be defined – or limited – by our interests as consumers" (2001: 477). Nevertheless ethical consumption is proving to be a pervasive agenda that has a wide resonance, but it is important to note that engaging in social action through ethical consumption may only be affordable to those who can pay a premium. The study tours to Cuba with NGOs, for example, are not inexpensive and are typically frequented by older and professionally successful people who have the resources to consume this style of tourism.

Another way to examine new moral tourism that is closely linked to ethical consumption is as a form of lifestyle politics which implies political solutions can be found at the individual level (Butcher 2003). Central to the debate on sustainability is the idea that individual lifestyle, what we consume and how we relate to other people are issues with potential impact on the wider world and on access to resources for coming generations. Tourism choices are strongly associated with image and lifestyle and hence a moral dimension to lifestyle is key to new moral tourism. Giddens (1991) refers to life politics as an attempt by people to reposition them culturally trying to make a difference both at an individual and broader level. Life politics, he argues, is a reconfiguration of the relationship of the individual to their society; consequently identity becomes a site of political change. For people participating in new moral tourism, this lifestyle choice in particular reflects their identity since they may derive a sense of self-actualisation, self-enrichment, self-expression, and development of self image (Hall and Weiler 1992: 8-9). Indeed for some, not unlike a pilgrimage "a journey makes sense as a coming to consciousness" (Clifford 1988: 167). Participation on study tours, for example, leads to increased awareness of certain issues and thus a realisation of self-image.

Attaching a universalistic moral framework to social practices as diverse as tourism is complex. How do we precisely assess what is good and what is bad, or what is appropriate or inappropriate in a context of relative values and cultural norms? Nevertheless a moral framework is important because it paves the way for a more nuanced understanding of the social changes that inform tourism practices and has implications for tourism as a development tool. A moral framework is clearly a facet of this convergence between tourism and development. To ignore it is to miss an important concern for subaltern rights. Arguably, 'new tourists' consider how tourism has the capacity to impact on people's rights. It is this concern that underpins people's choices to embrace new moral tourism and reproduces aspects of well-being and dignity in rights-based development. To interrogate this notion of a moral framework further, examination of NGO study tours to Cuba offers insights into what could be labelled 'rights-based tourism'.

The Moralisation of Tourism: Development-oriented Tours in Cuba

In order to lodge the discussion more closely in the micro-context, I now turn to the two organisations that are the focus of the field-based element of this book. The following discussion focuses on two NGOs and their study tours. It provides examples of how their goals and objectives demonstrate the values and characteristics associated with the new moral tourism paradigm and are an example of the ways in which development and tourism are converging. A comprehensive description of the organisations is provided. This section is important for two main reasons. To start with, very little has been written about the experience of partaking in development-oriented NGO study tours. Second, it is important for my subsequent analysis of participant experience, that the organisations, their agendas, their development aspirations, and the nature of the tours, be carefully described.

For several decades, development and tourism in developing countries have existed side by side yet have functioned largely independently of one another. To some extent, many NGOs have until recently, rejected tourism in developing countries. This could be because NGOs are sensitive to the negative impacts of tourism development. Development NGOs and human rights organisations often work with programs and initiatives focusing on issues such as displaced people, forced labour, women's labour issues, sexual abuse, craft production and so on, all of which can result from tourism development. Development NGOs then, have often been inextricably linked to the effects of tourism but it is only recently that NGOs have begun to consider tourism as a developmental tool. In 2001, the British NGO, Tourism Concern, commissioned a study into NGO involvement with tourism which indicated a very low level of NGO involvement in tourism.[7] This finding highlights the traditionally negative attitude of NGOs for tourism. NGOs have become increasingly aware, however, that tourism is growing rapidly throughout developing countries and thus needs to be integrated into development strategies in order to minimise its negative aspects, make it a fairer industry, and also provide positive vehicles for desired change. Diverse policies and strategies that demonstrate this range from pro-poor tourism initiatives, Cuba's development policy placing tourism at its forefront, and NGO study tours to developing countries.

Consequently, tour operators are increasingly establishing partnerships with local grassroots and international development NGOs so that their tours will incorporate visits to project sites which emphasise an educational agenda. Many of these organisations are ostensibly guided by a "giving something back" philosophy (Spencer 1999). In addition, some NGOs, such as those that are the focus of this research, are actively involved in tourism by organising study tours to the projects they fund abroad. This alternative form of tourism can be positioned squarely with various other alternative tourisms labelled ecotourism, cultural tourism, ethnic tourism, and edu-tourism. As discussed, new moral tourism perhaps is an appropriate label to encompass all these alternatives to mass tourism. As tours with NGOs have a strong educational focus and claim to offer meaningful experiences, parallels can readily be drawn with the experiences afforded to people participating in volunteer programs.[8]

To understand what NGO study tours are about and how they fit into the moral framework outlined above, it is necessary to describe the organisations behind them. Two organisations were selected to facilitate this research because of their commitment to development issues and responsible travel: *Community Aid Abroad Tours (OCAAT)*, a subsidiary body of the development agency Oxfam Community Aid Abroad (OCAA) located in Adelaide, Australia, and *Reality Tours (GERT)* a subsidiary arm of the human rights organisation Global Exchange located in San Francisco, USA. While both organisations demonstrate similar goals and operate in similar ways, it is interesting to note that *OCAAT* emerges from a development NGO background and *GERT* from a new social movement's background.

---

7   Kalish, A. (2001) *Tourism as Fair Trade: NGO Perspectives*. London: Tourism Concern.

8   Broad, S. (2001), Unpublished Doctoral Thesis looked at the role that volunteering played in the lives of volunteers at a gibbon rehabilitation project in Thailand. It also addressed volunteering as a tourism experience which provided both personal development and in depth experience of a culture, which were found also to be volunteers' main motivations for participation.

The way in which Oxfam Community Aid Abroad has embraced tourism is representative of a growing concern about NGOs and their efficacy. The climate is changing for NGOs as they go through a fundamental transition in terms of the need to stay relevant in the delivery of development assistance (Smilie 1997). NGOs have not only increased globally in number but have taken on new functions in an effort to stay alive in development (Fisher 1997). These changing functions typically stem from the growing criticism of top-down interventionist development and its failure to alleviate poverty. This in turn has stimulated many NGOs to search for alternative means by which to integrate individuals into global markets and to involve local populations in development projects (Fisher 1997). The incorporation of tourism into NGO activities is an example of this effort. Development has thus been a volatile industry embracing and then casting off a long series of ardent new strategies. It remains to be seen if the appropriation of various tourism strategies will be successful as a development tool.

On the other hand, human rights organisations like Global Exchange have emerged from the growing new social movements and are concerned with supporting marginalised and oppressed people in their acts of resistance against hegemonic forces. Collective or social movements are largely informed by a human rights discourse and are part of a 'new left', which is in contrast with 'old' movements associated with Marxism and class struggle. New social movements are positioned as social and cultural processes that represent subaltern voices, needs and desire (Parajuli 2001: 278). They often challenge the State, the dominant development paradigm and the subaltern status in an effort to internationalise the embodied realities and objectives of the indigenous actors of resistance (Guha 1983, 1988, 1996; Spivak 1988; 1996). NGO study tours are an important means by which new social movements internationalise subaltern issues in political climates that threaten their livelihoods and social and personal autonomy. In this way, NGOs encourage solidarity in the international community's effort to bring awareness to human rights issues.

In order to explore 'rights-based tourism' I selected both tour operators because their affiliation with NGOs would afford specific insights into the development-tourism nexus. Cuba provided a location where social development offered a peculiar blend of success and failure. Both Oxfam Community Aid Abroad and Global Exchange aim to make people more aware of international development and human rights issues. They do this through a range of fundraising campaigns and by organising educational delegations, which are intended to provide people with the chance to take part in 'responsible' tourism. The study tours aim to provide people with opportunities to meet local people and grassroots organisations and discuss local development issues, exchange ideas and establish meaningful relationships with people from other countries as well as with fellow tour participants. These are the precise elements that the 'new tourist' engaging a moral consumption wants to experience. My intention in the next section is to give a clear picture of how the two NGO tour operators and their philosophies fit into the 'new tourism' paradigm and a moral framework.

Oxfam Community Aid Abroad and Community Aid Abroad Tours

Oxfam Community Aid Abroad (OCAA), an independent, Australian, secular, not-for-profit, non-government, community-based aid and development organisation began in Australia in 1953 as a church-affiliated group. Since 1995, Community Aid Abroad had been part of the Oxfam International family – an affiliation of 12 Oxfams around the world.[9] At the time of writing, Oxfam Community Aid Abroad operated in more than 30 developing countries and in Indigenous Australia supporting long-term initiatives among poorer communities. It was stated by OCAA that its:

> Vision is of a fair world in which people control their own lives, their basic rights are achieved and the environment is sustained. They aim to increase the number of people who have a sustainable livelihood, access to social services, an effective voice in decisions, safety from conflict and disaster, and equal rights and status. Their work is a partnership through which Australians enable poor and marginalised people to control their own development, achieve equitable treatment, exercise their basic rights, and ensure the environment is healthy and sustainable. (*OCAA* website 2004)

OCAA's activities were aimed at overcoming the root causes of poverty and injustice by initiating long-term development projects, emergency relief, fair-trading, fundraising campaigns, and responsible travel. It was this responsible travel program that forms the key focus of this study. *Oxfam Community Aid Abroad Tours* was a voluntary not-for-profit special-interest group that supported the work of OCAA through community-based tourism. CAA Trading, a subsidiary of OCAA which incorporated shops and fair trade, had a department called *One World Travel* based in Melbourne. As a travel agency, it held the licence to operate tours, thus *One World Tours*, then *OCAA Tours*, operated under the travel agency although very independently.[10] In 2001, when I commenced the research in Cuba,

---

9   Oxfam International is an alliance of 12 non-government organisations working together in more than 100 countries to find lasting solutions to poverty. To achieve the maximum impact on poverty, Oxfam's link up their work on development programs, humanitarian response, lobbying for policy changes at national and global level. They seek to help people organise so that they might gain better access to the opportunities they need to improve their livelihoods. They also work with people affected by humanitarian disasters, focussing on preventive measures, preparedness, as well as emergency relief. The Oxfam approach is to work primarily through local accountable partner organisations, seeking to strengthen their empowerment. In Cuba, for example, Oxfam Canada has worked with grassroots organisations in community development projects focused on the transformation of Cuba's agriculture.

10   Name changes went from Our World Travel/Tours and then to CAA Tours and then to OCAA Tours.

OCAA Tours (*OCAAT*) were operating four study tours focusing on Aboriginal Australia. These included tours to Broome and the Dampier Peninsular, Lake Eyre and the Oodnadatta Track, the Kimberley, and The Red Centre. Other countries to which *OCAAT* were taking study tours to included Cuba, Guatemala, India, Laos, Madagascar, Solomon Islands, Tibet, Vietnam, Zambia and Malawi. *OCAAT* also offered tourists the option of designing their own tour for groups of six or more.

By putting people, their culture and environment first, *OCAAT* claimed to support responsible and sustainable tourism. The tours brought small groups of like-minded people together to learn about different cultures and provided opportunities to meet with local grassroots organisations and visit community projects (some supported by Oxfam). The economic benefits of travel were to be shared with local communities by working with local tourism initiatives and paying fair prices for modest accommodation, local guides, restaurants, entertainment and transport. Proceeds from all tours went to Oxfam Community Aid Abroad community development projects (*OCAAT* website 2002). These values and characteristics are clearly associated with the growth of a new moral tourism.

The tours combined educational seminars and project visits with leisure. *OCAAT* claimed to have pioneered this form of travel. It began operating tours to India in 1963 which provided tour participants with opportunities to witness the cultural and environmental impacts of tourism. They worked closely with communities to develop a sustainable tourism that offered the opportunity to visit "communities in a spirit of mutual understanding and reconciliation" (*OCAAT* website 2002). The underlying aim was to encourage responsible tourism in order to achieve a greater understanding of the world and the forces that shape it. *OCAAT* sought to raise tour participants' awareness by increasing their understanding of the issues pertaining to the culture they were visiting. In this way, misconceptions were broken down. This was particularly important in the case of Cuba, as the US government had disseminated ongoing negative propaganda throughout the decades of the Castro led revolution. Guiding principles of responsible travel, according to *OCAATs* philosophy, included understanding the culture being visited, respecting and learning from the people who were hosting the visit and treading lightly on their environment through small group tours.

The incorporation of education into alternative travel has grown rapidly in recent years (Weiler 1991), and new forms of tourism provide opportunities to learn about culture, development, conservation and so on (Poon 1989). *OCAAT* placed a strong emphasis on education. *OCAAT* advocated that travellers teach themselves and form their own opinions based on experience. A fundamental manner in which *OCAAT* facilitated learning was by providing tour participants with extensive and diverse pre-tour literature.

OCAAT believed that providing tour participants with educational literature prior to the tour, leads to a well-informed tourist, which in turn will lead to fewer negative impacts on the cultures visited. In the case of Cuba, the process of providing in-depth up-to-date information ensured that people generally joined the tour informed and ready to engage in discussion with the grassroots organisations with which they met. The Tour Notes stated that *OCAAT* provided a learning and leisure tour with a planned program to help understand Cuba from a human development perspective in a way available to few others. This focus supported the claims in the literature that 'new tourism' provides opportunities that are more specialised and targeted to individual interests (Poon 1993; Urry 1990).

Tour participants were asked to "remember that Cubans are generally well educated and very proud of their achievements especially in the areas of health and education where they lead much of the world, including parts of Australia and the US. This is one country where the generally run down nature of infrastructure including housing is highly deceptive" (Cuba Tour Notes). The tourists were prepared for an experience that allowed them increased opportunities to engage in dialogue with local Cuban people about issues pertinent to their daily experiences.

Pre-tour information for Cuba was extensive and included Tour Notes, a Resource Pack and a Responsible Travel Guide. The Tour Notes informed participants of the modest nature of accommodation, meals and transport, culture shock, attitudes to time and efficiency, and 'roughing it'. Advice was presented to prepare tour participants to be responsible tourists in matters such as dress, gift-giving, dealing with beggars, and photography and was aimed at maximising enjoyment and minimising participants' impact on host culture. Tour Notes did not, for example, advocate gift-giving to individuals as this may create jealousy and false expectations. Participants were advised that it is more appropriate to bring items, such as equipment for local schools (drawing materials, exercise books and other books, pens, pencils), churches, community centres or health clinic (basic medicines and first aid supplies). *OCAAT* also recommended that participants approach photography with circumspection and that they send photographs back to the school or community, as a means of saying "thank you" for the school or community's hospitality. Several pages of basic Spanish language hints were provided to assist tour participants with pronunciation and provided useful common words and phrases.

The Resource Pack was an extensive kit aimed at providing tour participants with a wealth of reading material to review before departing for Cuba. *Travel Wise and be Welcome*, an *OCAAT* publication was included in the pre-tour package. It aimed to educate travellers about the positive and negative impacts of tourism in developing countries on environment, indigenous people, women and children. The booklet explained that the costs of mass tourism can be reduced considerably through responsible travel.

Global Exchange and Reality Tours

Global Exchange is a San Francisco based non-profit human rights organisation operating as a research, education and action centre, which works for global political, economic, environmental and social justice, and operates within a new social movement's framework. The founders of the organisation were a nutritionist, a welder and an anthropologist who set out to create opportunities for the establishment of global relationships. Since Global Exchange's inception in 1988, they have worked to increase public awareness in the US of global issues while building progressive grassroots international partnerships. The goals of Global Exchange are: to educate the US public about critical global issues; to promote respect for the Universal Declaration of Human Rights; to encourage democratic and sustainable development; and to link people in the US with people in the global South who are working for political, social and environmental justice.

Global Exchange pursues these goals through five programmatic areas that seek fundamental change in US and corporate policies. The five program areas include political and civil rights campaigns, economic rights campaigns, fair trade, public education and *Reality Tours*.[11]

Global Exchange's *Reality Tours (GERT)* aims to educate the public about domestic and international issues through socially responsible small group tours. Participants in these intensive educational delegations examine political, economic and social trends in a number of countries, including Cuba, South Africa, Mexico and Vietnam. *GERT* also hosts human rights delegations to observe and report on events in areas of conflict, and serve on election monitoring delegations. Through direct contact with grassroots leaders, educators, political actors, artists and indigenous cultures, *GERT* hopes to give individuals concrete activism tools for use when they return home.

Like *OCAAT*, *GERT* promoted its travel as having specific educational characteristics that included a focus on development and human rights issues, authenticity, opportunities to meet with like-minded people and opportunities to contribute resources to aid the NGOs' activities in the target country. These values and characteristics were all indicative of a new moral tourism as outlined in the literature (Brown 1998; Cohen 1987; Hinch and Butler 1996; Lash and Urry 1994; Poon 1993; Weiler 1991).

Strong emphasis is placed on the principles of experiential education; it was the notion that travel could be educational and positively influence international affairs

---

11 On 13 April 2009, the Obama Administration lifted travel restrictions to Cuba for US citizens as part of the economic trade embargo. The travel restrictions associated with the half century long embargo is discussed in the next chapter, along with the provisions in the laws that enabled US citizens to travel there. Since the late 1980s *GERT* has taken thousands of US citizens on educational trips hosted by the Cuban Institute for Friendship of the People (ICAP). Information about the legality of travel to Cuba is supplied on *GERT*'s website and all promotional material associated with their Cuba tours. Global Exchange is a Travel Service Provider licensed by the Office of Foreign Assets Control (OFAC) of the US Department of the Treasury to organise educational and people-to-people exchanges in Cuba. This means that when US citizens travel with GERT to Cuba they are travelling legally. Thousands of US citizens travel to Cuba every year independently and without a license (therefore 'illegally') because they believe their government does not have the right to dictate which countries they can travel to.

that motivated the first *GERT* tour. As part of the growing new social movements oriented towards human rights and transnational in operation, *GERT* utilises tourism as a means of educating tourists to become more active in campaigning for international human rights. Their tours aim to teach about how people in the West, both individually and collectively, contribute to global problems. The tours obviously cannot provide immediate solutions or remedy the world's most intractable problems, but Global Exchange argues that the tours are not simply a brand of voyeurism either. *GERT* argues that by joining one of their delegations or tours, a participant will have the chance to learn about unfamiliar cultures, meet with people from various walks of life, and establish meaningful relationships with them. Most significantly, *GERT* claims to provide participants with a new stance from which to view and affect US foreign policy. It also hopes to prompt participants to examine related issues in their own community and society.

Alternative educational travel is promoted by *GERT* as a way of replacing apathy with deeper understanding and a sense of empowerment in which relationship building is essential. Thus every tour seeks to establish a global connectedness through people-to-people ties by introducing participants to individuals and communities that most travellers would not necessarily meet independently. These ties, in many cases, are the result of symbiotic programs between local communities and Global Exchange. *GERT* aims to facilitate new relationships between these communities and their tour participants, be they individuals, universities, or membership associations. The tours attempt to provide opportunities to experience authentic interactions with the community, the authentic 'backstage' that MacCannell (1976) talks of.

GERT, like *OCAAT,* offered pre-tour resource information in order to facilitate the preparation and education of the tour participants and in this way encouraged tour participants to educate themselves about their destination and the issues that would be addressed on the trip before tours departed. For Cuba, each participant received a general information pack, which was tailored to the particular tour. Generally, the pack included a letter from the *GERT* Director stating the license details to be presented to immigration on return to the US, the itinerary, recommended readings, a code of ethics for tourists along with a booklet of 301 tips for socially responsible travellers, specific information advising tour participants about what to expect at immigration on their return from Cuba, and an Orientation Pamphlet that included a range of pragmatic information on Cuba and socially responsible travel.

This socially responsible travel information is one of the key ways *GERT* prepared its participants for the study tours. For example, the 2001 Women's Delegation pre-trip Orientation Pamphlet advised participants to overcome the 'Ugly American' syndrome by observing some fundamental points that included:

> make a list of what you are expecting to find before you go ... This may aid
> you in understanding what ideas and stereotypes are in the minds of people
> you talk to on your return as well as being a way for participants to measure

their growth and learning. Americans can, at times, be somewhat boorish and arrogant tourists. Some ways to help fight that image: become acquainted with local customs and respect them. Do not make promises to local people or to new friends that you cannot keep ... Even if you do make a promise of aid and follow through on it, a shotgun approach of helping whoever you meet is not always the best answer. We especially discourage giving money or material aid to individuals.

In order to facilitate socially responsible travel, participants in *GERT* tours were provided with a *Current Issues and Background Reader*. As part of its education agenda, the organisation encouraged its participants to develop a learning methodology to overcome the possibility of participants being overloaded with new information. NGOs, such as *OCAAT* and *GERT*, provide extensive pre-tour material to tourists in an effort to break the "authority of pre-tour narratives as constructed by the tourism industry and embedded in Western discourse" (Bruner 2005: 22). Generally, Western touristic discourse presents Cuba as an exotic island paradise. The two NGOs in this research provide their tour participants with informed and diverse selection of materials in order to transcend this trite touristic image of Cuba.

GERT found that once people return home they were anxious to share the knowledge they had gained with others; it therefore encouraged participants to set up speaking engagements for their return or to ask if there is any specific information they can gather for organisations. One of the purported benefits of a *Reality Tour* was that participants returned home with an enlightened perspective, an expanded world view, and a sense of how they each could communicate what they had learned to others. In some ways, this ties into the sense of *flaneur* and cosmopolitanism, where one leaves home to be educated through travel and achieves competence in cultures, which are initially alien (Hannerz 1996) and encourages a sense of moral superiority associated with new moral tourism. *GERT* also sent the participants contact lists of the groups with whom they met, so the participants could stay involved after their return home.

One of the stated goals of *Reality Tours* is that the people in the destination country benefit from the tour experience. One way this was achieved, was to provide pre-packaged medical supplies; participants were encouraged to drop by the Global Exchange office and pick up a box of aid prior to the departure of their tour. Many past participants have donated time, technology, medicines, books, and money to the groups they have met abroad. After travelling, other participants have supported co-operatives in destination countries by purchasing and importing local often hand-crafted products. In some cases local organisations and individuals benefited from the publicity, new local interest, and the renewed local support the tours have generated. Most people in the groups with whom I worked brought some form of aid, with the majority contributing first aid kits and primary school resources.

The tours operated by these two NGOs are very similar in their styles of operation, yet their aims are different. Global Exchange very specifically seeks to educate the American public about the effects of US foreign policy on poor countries. While Oxfam aims to increase awareness of international development efforts with a view to encouraging support for their aid campaigns. Both organisations utilise tourism as an educational tool as part of their programmatic approach to human rights and development.

The establishment of a moral theoretical framework is useful for investigating the development-tourism nexus in Cuba as it offers avenues to question 'new tourism' and its instrumental effects. A moral imperative drives an increasing connection between global social movements, consumer engagement in social change and endogenous development endeavours. New tourism cannot be seen as some benign form of tourism. NGO study tourists are actively engaging with local people to learn about economic, social and cultural development. Part 3, 'Rights-based Tourists in Cuba' addresses tourists' particular motivations for participating in 'new tourism', and explores the educative role these tours play. The twin lenses of development theory and tourism studies are used to scrutinise the NGO study tours taking place in Cuba and to inform a broader understanding of the complex issues underpinning globalised networks of social change.

PART 2
On the Ground:
Cuba, Social Development, and NGO Study Tours

Chapter 3

Social Development in Revolutionary Cuba

Cuba is a country famed for charming beauty and therefore attracts many tourists. It is also a poor country suffering the combined effects of the collapse of the Soviet Union and the constraints of the US embargo. At the end of the twentieth century, *The Buena Vista Social Club* movie wove the biographies of forgotten musicians into a portrait of modern Cuba in a mode of remembrance. The film revitalised and projected images of Havana around the world. These were nostalgic images: 1950s Chevrolets being driven down streets lined with old Spanish colonial mansions – their faded splendour crumbling; people salsa dancing at one of Hemmingway's local haunts, smoking cigars and drinking Havana rum. The images portray the Cuba most tourists expect to encounter. Some tourists, however, visit Cuba to learn about the development challenges and accomplishments of its tenacious communist revolution. It is an island which has experienced some wonderful successes, learned some hard lessons and faced many obstacles. NGO study tours focus on social development, teaching visitors about the merits, failures and challenges of Cuban socialism. In this age of globalisation and neoliberalism Cuba offers insights into alternative models of development. This chapter details Cuba's model of development, the US-imposed embargo and its effects, the 'Special Period', and the subsequent economic reforms and social changes. It is important to understand Cuba's economic crisis in order to appreciate the nature of its social development and participation by local people.

Cuba banished hunger in the early 1990's but malnutrition returned suddenly. People were forced to lower their standards of living; recycling and inventiveness became almost symbolic of Cuban national identity. In an increasingly impoverished society, everything was saved and nothing was wasted. People endured long blackouts in sweltering heat and if one lived in an apartment building, as do most Cubans, a morning shower had to be carried up stairs in buckets. Public transport became almost non-existent. Many Cubans rose in darkness and walked for long hours to school and work – offices, fields and factories. People would leave work early in search of food for the evening meal. They cooked with wood, often chopped up furniture, because there was no gas or kerosene. "*No es facil*", "it's not easy" was a favourite phrase, still heard today, and somehow through it all most people, certainly the ones I met with, managed to retain their sense of humour and dignity of very proud people as they sat on the stoop at night having conversations and offered *agua con azucar*, sugared water, to their visitors.

This was what Cuba struggled with after the loss of trade and subsidies following the 1989 collapse of the Soviet Union. The government dubbed the period between 1990 and 1994 'a Special Period in Peacetime' in recognition of the 35 per cent decline in GDP. Although growth resumed in 1994, the economy has never fully recovered to pre-1989 levels of GDP. At the same time, the US government tightened the economic blockade against Cuba. Washington's intention was that the Cuban people, driven by hunger and despair, would rise up against what it termed 'the repressive Castro dictatorship'; the US government sought its own revolution.

Many economic reforms and social changes in the years of the socialist revolution were due to two parallel crises, according to Juan Antonio Blanco, a leading Cuban intellectual (Benjamin 1997: 120-121). The first crisis was structural; the government abandoned the development model imported by the Soviet Union and designed a model suited to Cuba's Latin American context. The government subsequently designed a more effective strategy abandoning the development model imported from the Soviet Union. The second crisis stemmed from external factors, such as the collapse of the Soviet Union and the intensification and extraterritorial extension of the US embargo. These crises led to the adoption of new economic measures and Cuba's reintegration into international economic markets. Most relevant to my analysis, the reintroduction of international tourism now provides one of Cuba's main sources of income while the country faces the challenges of the globalised world economy, while still (at the time of writing) subjected to the ongoing trade embargo imposed by the United States.

Cuba's economic crisis coloured everything to do with everyday life and social development. *El Periodo Especial* had devastating effects on both the economy and people. The Special Period refers to the economic crisis during the 1990s instigated by the collapse of the Soviet Union and the tightening of the United States trade embargo on Cuba. Cubans prefer the term 'economic blockade' because the United States interfered in Cuba's trade relations with other countries. Since 1990, the country has faced serious economic constraints resulting firstly from the disintegration of communism in Europe, particularly the dissolution of Cuba's Soviet trading partner, and the subsequent disruption of trade relations. This impacted heavily on Cuba's economy, which promptly collapsed. From 1989 to 1993, as mentioned, gross domestic product declined by 35 per cent, import capacity plunged by 75 per cent and the deficit reached 33 per cent of gross domestic product (Gordon 1997). At the outset of the revolution Cuba had forged economic ties with other communist countries importing medicine, food, fuel and heavy equipment for mining and agriculture. With the collapse of communism in Eastern Europe and the Soviet Union, those ties were lost. Subsequently, the dramatic decline in the availability of food, clothing, construction materials, and medicines,[1] deeply altered the daily lives of Cuban people. In addition, the government was forced to stop investing in infrastructure. Suddenly, there were

_____

1   Throughout the 90s the enforcement of the blockade had a significant impact on Cuba's health care. For example, a Swedish corporation prohibited for selling medical equipment to Cuba because it contains a filter patented under US law. Spare parts for x-ray machines from France, neurological diagnostic equipment from Japan, parts to clean

no imported fertilisers, pesticides, petroleum or parts for machinery. As a result, the production of sugar cane, the source of about 30 per cent of its hard currency, plummeted (Sinclair and Thompson 2001). In 1994, the worst year of the Special Period, the island's agriculture was producing only 55 per cent of its 1990 output (Sinclair and Thompson 2001). A lack of materials also impacted on the other main sources of hard currency: the nickel and tobacco industries. These combined factors led Cuba to place tourism at the forefront of development policy.

The US exacerbated the situation by tightening its trade embargo to exclude Cuba from international financial markets. The underlying purpose of the embargo had always been to impose a capitalist regime on Cuba and destabilise the government by creating hardship among the Cuban people. Ships that deliver goods to Cuba were prohibited from docking in US ports for the preceding six months. In 1992 the US Congress took the opportunity afforded by the collapse of the Soviet Union to introduce the Torricelli Act. It mandated sanctions against foreign subsidiaries of US companies that trade with Cuba and against foreign nations that engage in commerce with the Cuban government. The Act was intended to force Cubans to overthrow their government. When Cuba opened up to some foreign investment in 1996, the US Congress passed the Helms Burton law to further deprive people of critical imports of food and medicine from *other* countries and level sanctions against US citizens who travelled to Cuba without US permission.[2]

The cumulative effect of these two Acts was devastating for the Cuban people. Economic difficulties led to serious deterioration of health and social services, particularly in urban areas like Central Havana[3] (Campbell 2001). Not only did agricultural production come to a standstill causing serious food shortages, public transportation almost ceased, and petrol supplies for private cars were extremely limited. Electricity was closely rationed, and long blackouts were common in the city. According to Uriarte (2002), as Cuba attempted to cope with the economic crisis in the 1990s the government struggled to maintain a financial commitment to the values of universality and responsibility for social benefits. Indeed there was much international speculation as to whether the universal social entitlements the government had instituted would survive such a profound crisis.

---

dialysis machines from Argentina, Italian made chemicals for water treatment and many others were also prevented by US law (Gordon 1997).

2   Recently the Obama Administration eased these travel restrictions to Cuba for US citizens, but little else has changed regarding foreign policy towards Cuba.

3   Inner city Havana, called Centro Habana, was founded over 450 years ago and is Cuba's most densely populated municipality with about 170,000 people living in just over three square kilometres of tenements. It suffers from poor housing and prevalent communicable diseases, with most people lacking basic sanitation and safe drinking water (Campbell 2001). The municipality started a community intervention program in 1996 aimed at improving the urban environment through better housing, water supply and the development of community centres (IDRC: 2003).

Cuba's model of social development aims to provide universal access and equity across society. The government is responsible for the funding and delivery of social entitlements. This model of social development had, since the revolution, instituted free health care and education, provided a universal government-sponsored pension system, and provided a safety net of benefits, which included food subsidies, protection of workers' employment, housing and assistance to vulnerable families. In practice, this model enabled Cuba to increase adult literacy to 100 per cent,[4] decrease infant mortality,[5] minimise drug use and crime compared with other Latin American countries, and effectively end homelessness.[6] The model has been highly commended internationally. For example, the former Secretary General of the United Nations, Kofi Annan (11 April 2000)[7] endorsed Cuba's development in these terms:

> Cuba's achievements in social development are impressive given the size of its gross domestic product per capita. As the human development index of the United Nations makes clear year after year, Cuba should be the envy of many other nations, ostensibly far richer. [Cuba] demonstrates how much nations can do with the resources they have if they focus on the right priorities – health, education and literacy.

Following the publication of the World Bank's 2001 edition of World Development Indicators,[8] the World Bank President, James Wolfensohn, hailed free healthcare and free education in Cuba as model investments (Lobe 2001). At the time, the World Bank commended the Socialist government of Fidel Castro for providing for the social welfare of the population and conceded that Cuban socialism has succeeded where the neoliberal ideology of the World Bank and International Monetary Fund has manifestly failed elsewhere in Latin America (Lobe 2001).

From the outset, Cuba's model of development aimed to link economic growth with social justice by pairing economic changes with social initiatives. One of the best examples of this is the agrarian reform initiated at the beginning of the revolution. The reforms ended *latifundia* (large property ownership) and distributed land to thousands of landless peasants. Other examples of fundamental reform that took place at the outset of the revolution included programs that provided healthcare and education in rural and urban areas. A national health system with rural health services was one of the first initiatives. Another initiative was the National Literacy Campaign of 1961. It mobilised teachers, workers and

---

4   http://www.unicef.org/infobycountry/cuba_statistics.html.

5   Oficina Nacional de Estadisticas 2008.

6   Oxfam America.

7   Kofi Annan, UN Secretary General addressing the opening of the first South Summit in Havana.

8   World Development Indicators, 2001.

secondary school students to teach more than 700,000 people, primarily farmers how to read. In only one year the illiteracy rate was reduced from 23 per cent to 4 per cent in only one year.[9] Urban reform focused on halving rents and providing opportunities for tenants to own their own homes; there was in addition a program of housing construction for those living in marginal shantytowns. New housing and measures to create jobs and reduce unemployment, especially among women, rapidly transformed the former shantytowns.[10]

These successful social programmes are the focus of NGO study tours which expose tourists to carefully selected groups and sights. For the Cuban government, international NGO study tourists are seen as a means to build favourable international relations as tourists can give 'first-hand' accounts of the country and its economic and social achievements.

Cuban social policy is characterised by its emphasis on universal access to education, health and social benefits. This is identified as part of a 'social wage' that workers accrue in addition to their monetary wages. Cuba's social policy embraces the equitable distribution of benefits across all sectors of the population, sometimes favouring the most vulnerable. During the first 40 years of the revolution, the government significantly reduced the differences in incomes between the lowest and highest paid. Women benefited from education opportunities and entered the labour force in large numbers.[11] Inequities between different ethnic groups have also been reduced. The programs and subsidies that made up Cuba's social welfare aimed to cover people from cradle to grave. The 1999 Human Development Index (HDI), ranked Cuba 51 out of 177 countries based on longevity, education and a decent standard of living.[12]

Because the revolution had already made such advances in its first three decades, *el Periodo Especial* had very dramatic effects on people. Food production fell drastically and food available on *la libreta* (the ration card) decreased too. People had to go without or pay black market prices. Cubans often recount that there were times when they went to bed hungry at night. The 2001 United Nations Human Development Report states that 19 per cent of the population was undernourished between the years of 1996 and 1998, and this was a time when the economy was starting to recover. Health problems associated with malnutrition developed among the elderly, rural and those living in greater urban poverty. Special nutritional programs were established to rectify the nutritional deficits of expectant mothers, day hospitals for pregnant women were established to care for the most at-risk cases, meals were provided in workers' cafeterias, and the government attempted to control the spread of HIV/AIDS by institutionalising people who were HIV

---

9   Centro de la Economica Mundial, 2000.

10   GDIC – Grupo para el Desarrollo Integral de la Capital, 2001.

11   United Nations (2003) UNFPA Country Profile Cuba, http://www.unfpa.org/profile/cuba.cfm.

12   UNDP 2007/2008 Human Development Report.

positive. By 1995 health conditions were overly improved, and infant mortality astoundingly had not been affected.[13]

Cubans have demonstrated capacity for innovation and transformation, which is most noticeable since the beginning of the Special Period. High levels of education and a propensity for adaptation and invention in the face of huge constraints are evident in the ways Cuba has transformed the economy: from the urban agriculture movement,[14] the development of environmentally sound agricultural practices[15] and the application of alternative energy sources.[16] The Special Period accelerated the rate at which prior concepts and styles of government were becoming obsolete. Even as they proved effective in the past, they were becoming increasingly inappropriate and even counterproductive (Benjamin 1997: 124). The early years of the 1990s focused on the economic crisis and the need for innovative measures to overcome it. The second half of the 1990s focused on responding to emerging negative impacts of those measures (Benjamin 1997). I explain these measures and subsequent impacts below.

During the 1980s the World Bank and International Monetary Fund imposed 'structural adjustment' throughout most of Latin America. The 1980s saw the external debt of the Third World increase enormously. Many countries were pressured by the IMF and the World Bank to implement structural adjustment programs, thereby selling profitable State enterprises while cutting expenditures on social services, with profits and savings paid into what Latin Americans call *la deuda impagable*, the unpayable debt. If Cuba had succumbed to structural adjustment, it would expect to be in the same position as other Latin American countries with widespread poverty, unemployment and violence. Instead the levels of poverty, violence and unemployment are low. This is not to say that Cuba does not have poverty. Rather, Cuba has restructured its economy while maintaining its commitment to its socialist values.

In contrast, the Cuban government implemented a temporary structural adjustment approach that permitted self-employment, raised prices, legalized the US dollar, and decreased government subsidies of State enterprises. It further instituted

---

13   UN Human Development Report, 2001.

14   As Cuba continues its efforts to find new ways to improve its capacity to feed its population, a very successful alternative that has emerged over the last decade is that of increasing food production in urban empty spaces. This approach has created a model of food security for the rest of the world (United Nations 2002). Many city dwellers have started growing their own food to supplement their diets and incomes. The Cuban government is supporting these individual efforts with a food program that encourages urban agriculture. IDRC, the Canadian International Development Agency (CIDA), and Oxfam are working with the Cuban NGO Fundacion de la Naturaleza y el Hombre, to assess the longterm potential of urban agriculture and to disseminate technologies for growing food in confined spaces (IDRC 2003).

15   The development and use of organic fertilisers and pesticides.

16   CubaSolar is an NGO paving the way for solar energy, particularly at schools and homes in rural areas.

a series of stabilisation measures to restructure the economy. These measures proved highly successful and recovery commenced in 1994 with growth trends averaging 4.3 per cent annually until 2000. During this same period, most of the economies of Latin America were showing either a flat or negative growth trend (United Nations 2003). These measures represented new conditions for foreign investment, addressed critical problems in the internal finances of Cuba and opened up new economic opportunities for Cubans to work privately (Uriarte 2002).

Although international assistance to Cuba is limited, as mentioned, the Cuban government has avoided plunging into debt to relieve its economic situation. Rather, Cuba has accepted assistance from various United Nations programs[17] and international non-governmental organisations (NGOs), for example, the Canadian organisation International Development Research Centre (IDRC) and Oxfam Canada, Belgium and the US[18] among others. The Cuban development strategy contrasts sharply with those of Latin American countries and other developing nations, where foreign corporations invest precisely because there are little or no environmental restrictions and protections for labour as compared with First World countries. The Cuban government's policy requires that foreign nationals include provisions for waste disposal and land use consistent with sustainable development in their investment proposals (Gordon 1997).

Since the beginning of the revolution the Cuban economy has been centrally controlled. Government owned subsidiaries have produced goods and services, and the government has been the single importer and the single employer. But with the economic crisis of the 1990s, Cuba needed measures to encourage foreign investment and private domestic activity in some sectors of the economy. In 1989, the government became concerned with making changes in response to the specific conditions in Cuba while respecting the principle of 'self-determination', one of the golden rules of Marxism-Leninism. This entailed Cuba (the government) to be at liberty to act as it saw fit. Hence, Castro argued that all people and parties have the right to dialectically interpret revolutionary theory according to their specific circumstances. Further, he argued that Cuba has the right to reconstruct capitalism within its national borders as it sees fit. Just as any capitalist country has the right to build socialism within its national borders (Suarez Salazar 1999). It should be noted, however, that the often harsh responses by the government to dissident activities indicates that in reality people do not have the right to self-determination if their views or activities contradict the espoused ideology of the government.

---

17   United Nations assistance has played a facilitating role for development in Cuba. UNDP programs have assisted in technology and industrial development for 25 years. UNDP's role has since shifted to include social, economic, financial and energy areas to help Cuba cope with the difficulties of the Special Period, supporting the Cuban government in its efforts of decentralisation, equity, strengthening of gender thematic work, improvement of the coverage of social services and the strengthening of the local economies (United Nations 2003).

18   The embargo does not extend to international NGOs such as Oxfam America.

During the Special Period, the government went into partnership with certain Canadian and European companies, notably opening doors to tourism, which was to become the main source of hard currency for the economy. In 1993, to support international investment, the government legalised the US dollar creating a dual economy; US dollars and Cuban pesos. The growth of foreign investment also presented people with job opportunities in hotels, offices, and services administered by international companies. As well Cuban corporations were created to provide services to these sectors providing ensuing employment (for example, Cubalse and Corporacion CIMEX). The advent of a dual economy created segregated commercial and labour markets, one operating in dollars and the other in pesos with very different pay systems and working conditions. Tourism operated in the dollar economy and was thus separated from the everyday life of most people unless they worked in tourism.

Two other significant reforms radically changed the socialist economy. In the early 1990s (post Soviet Union collapse), the State privatised large areas of agricultural production by turning former State owned farms into co-operatives called Basic Agricultural Production Units (UBPCs). The government opened the Farmers Market, for both private and State producers, selling in pesos with prices determined by supply and demand between farmers and consumers. This effectively provided desperately needed additional sources of food, but at very high prices. Also in 1993, the government introduced opportunities for self-employment permitting private enterprise (*cuentas propistas*) in a broad number of areas ranging from taxi drivers, hairdressers, fishermen, restaurant owners to video producers. To begin with, many people seized the opportunity for self-employment. When the government recognised how successful private enterprise was and that it successfully competed against State operations, high fixed taxes were introduced to discourage the growth of the industry. The combination of these measures improved the economy significantly with the GDP moving upward from the mid 1990s with the deficit narrowing and imports expanding. However, prominent Cuban dissident, Elizardo Sanchez Santacruz[19] notes that the heavy taxes imposed by the State drove many flourishing private businesses to close their doors, thus indicating that he believes the Castro led government will not let capitalism take hold (Solman 2001).

Market-based reforms immediately showed positive results. But even as the success of the measures was becoming evident, the government expressed reluctance at having to implement them. The measures were projected as temporary 'necessary evils' which would be reviewed once the crisis was over. Serious concerns were raised by those working in the development field in Cuba about the effects of the reforms on the people. Partial market reforms have introduced inequalities and the increased economic disparity has given rise to the emergence of new social problems, such as prostitution. As a result of a 'new economy' for

---

19   Elizardo Sanchez Santacruz is director of the Cuban Commission for Human Rights and National Reconciliation and was jailed for nine years for his politics.

the first time since the revolution social stratification has emerged. People have questioned whether the Cuban government could continue to encourage the free market and resist it at the same time and how it intended to face these challenges. Recently it appears that the government is restricting private enterprise through increasing taxes. This began in 2003, and in the following year US currency was banned. In 2004 the government began to replace US dollars with convertible pesos. The convertible peso established in the 1990s had the same value as the US dollar and would continue to be exchangeable with the dollar and other hard currencies (especially the EU) according to the international market rate.

The most critical effect of the reforms has been the increase in income inequality resulting from the transformation in the structure of the labour market. The crisis measures meant that the labour force privatised rapidly and many jobs were created in the tourism industry, in areas with heavy foreign investment, and in private national industries specifically created to service this sector. There was an increase in the numbers of self employed workers, independent farmers, and farmers working in co-operatives. The difference between conditions and rewards for those working in the State and in the new economy sector has varied greatly. Workers in the emerging new economy had access to technologies, office equipment and supplies and luxuries not available in State enterprises (for example, air conditioning). In addition to a regular salary in pesos, these firms have provided workers with benefits such as clothing and toiletries, which were hard to come by during the Special Period. Some of these enterprises even offered workers at least some of their salary in dollars. As a result, there has been a growing gulf between those with access to dollars and those without who are trapped in the peso system (Solman 2001).

Prior to *el Periodo Especial*, the highest paying careers were those of doctors and engineers. However, with the introduction of the new economy, a waiter in a tourist hotel, previously one of the lowest paid jobs in Cuba could obtain a set of rewards that included a salary in pesos, tips in dollars, benefits and improved working conditions. These were worth much more than the rewards accruing to top professionals who worked for the State and earned high peso salaries. This situation was commonly referred to in Cuba as the 'inverted pyramid', a phenomenon that reflected the devalued return on education and professional preparation in the new economy. As a result there was a mass exodus of public service workers into low-level service jobs in the tourism industry which access the rewards of the world economy (Uriarte 2002).

Perhaps the most salient factor contributing to the growing income inequality has been the unequal access to dollars that the new economy provides. Many Cubans have been earning dollars primarily through self-employment (*cuentas propistas*). Such work includes home restaurants, unofficial taxi drivers, and performing various services or selling goods on the black market. People also gain access to hard currency through remittances from family and friends living abroad, mainly in the United States. It has been estimated that around one billion dollars per year flow in to the country through remittances which has provided not only a major source of hard currency but also a major source of family income.[20] In many of the homes I visited, evidence of hard currency could be gauged by the appliances. The cracked oatmeal-coloured exteriors of apartment buildings with faded remnants of previous colours often belie the modern conveniences within. Families with access to hard currency may have modern stereos, televisions, fridges/freezers, fans, and washing machines.

Within any given family, different forms of participation in the 'new economy' can exist: work for the new government businesses, or work in private enterprise in the formal and informal economies. Young *jineteros*, hustlers, often earn more money in an afternoon showing a tourist around than the monthly income their fathers earn in a job with the State. Many older Cubans that I talked with complained about their children earning dollars (or convertible pesos since 2004) to purchase items only available in hard currency, even though they may have had limited access to dollars. Due to these circumstances, Cuban families were faced with new pressures, which led to considerable family stress. It may be that the rising divorce rate is an indicator of family stress (Federation of Cuban Women 2002, pers. comm. 20 May).

Another burden for families and government is anxiety about specific social problems associated with the growing tourist industry. Commercial sex workers and petty criminals have increased in cities with high tourist numbers (Benjamin 1997) a typical impact associated with tourism development. In an attempt to minimise the numbers of Cuban youth flooding into Havana to find employment in either the new economy or illicit activities, such as commercial sex work, legislation has been introduced to prevent people from moving away from their province. This is significant because the government claims to have almost eradicated prostitution and crime at the outset of the revolution. Sex shows were closed down immediately and prostitutes (*jiniteras*) were sent to trade schools to be re-educated in 'respectable' professions. By the mid 1990s, prostitution had become more visible and, Cuban women were subsequently, banned from the hotel rooms of international tourists. The Federation of Cuban Women (FMC) claimed that the commercial sex workers were well-educated women compared to the Batista[21] days, and that it is not only money that motivated them but a desire to access the glamour of international tourism. The FMC concluded that social problems resulting from the new economy, such as the increase in commercial sex workers, or *jiniterismo*, reflect a moral crisis amongst Cuban youth, rather than a response to the poverty of *el Periodo Especial* and perhaps this is because Cuban youth lack the 'revolutionary spirit' of their elders.

---

20 Centro de Investigaciones de la Economia Mundial (2000).

21 Fulgencio Batista staged a coup and became President in the early 1950s. It was during this period that gambling and prostitution proliferated with the opening of numerous casinos owned and operated by infamous US mafia members. Crime and poverty were rife during this period.

Juan Antonio Blanco asserted that the government has aimed to build on its ethical values of solidarity, independence and social justice at the same time that it has been forced to integrate itself into a global economic system (Benjamin 1997: 122 and 128). Indeed, the fundamental difference between Cuba and other countries undergoing liberalising reforms is the government's commitment and political will to protect its people from the most damaging effects of liberalisation and from the impact of the economic crisis. In the values clash the government claims to be maintaining values underpinning social policy: equity, universality, and government control (Uriarte 2002). The advisor to the Minister of Tourism, Miguel Figueras believes that while Castro has attempted to apply *some* of the formulas of capitalism to resolve its problems, these are essentially only methods of organising production (2002 pers. comm. 6 June). Most importantly, Cuba still has its benefits (albeit threatened) of social equality, universally free education and healthcare (2002 pers. comm. 6 June). Travelling through Cuba today, however, one sees that these reforms were not sufficient to prevent the lack of supplies for hospitals including medical equipment or for education including the exodus of school teachers to employment in the new economy.

The government remain committed to the key values underscoring social policy, and continue to assert that it will maintain its social development model. This is demonstrated by the commitment to equitable and universal access to all health and education services (and other benefits). In addition, the State still tries to command the economy by retaining its responsibility for funding, developing, and providing benefits and services, in spite of the move over the last decade towards decentralisation.

Thus far, the socialist government subsidised system has struggled to maintain itself but it is clear that the Cuban government has a strong political will to maintain its social safety net. Throughout this period, Cuba actually increased spending of the gross domestic product on social programs by 34 per cent.[22] The 1990s have led to complex social changes that now challenge the government's commitment to equity.

In order to continue to strive for positive social outcomes, Cuba must find new means to do so, as it faces an escalating set of social demands. In response, the government is not reducing services or privatising them, but instead attempting to transform the framework of services. According to Oxfam America, "as Cuba develops effectiveness that does not rely solely on the massive deployment of resources [something it lacks], it will need to use its great reservoir of experience in approaching problems preventatively" (2002: 4). The creative ways that Cuban people deal with financial crisis is precisely the focus of development-oriented tour groups, which this book explores.

The Cuban government is introducing new and more diverse initiatives underpinned by the increasing acknowledgement that universal policies and centralised initiatives alone could potentially overlook the needs of specific and

---

22   UNDP – Country Program Outline for Cuba 2003.

newly vulnerable people in this more diverse environment such as rural families, lowly paid civil servants and those who do not access hard currency. As social changes associated with new wealth has lead to a host of new problems, the government is devising more collaborative and integrated approaches towards maintaining its commitment to service delivery. For example, the local community development movement, established over the last decade presents a model of small scale, place-based, participatory planning and the monitoring of services, which could enhance the scope and effectiveness of service delivery. Thus far, useful methods developed at the community level include focusing on families and communities, collaboration and co-ordination of activities between local entities, connecting families and communities to enhanced community networks, developing the capacity of residents and local government to participate in local policy development and in the monitoring of local services (Uriarte 2002: 5).

Even as the government has maintained its social development model and continued funding of social services, certain aspects of the model are undergoing transformation to improve its reach and effectiveness. Although the revolution has set in place equity and universal accessibility, there has been a lack of efficiency in the delivery of services. Many social indicators have not yet reached levels achieved prior to the crisis in the late 1980s. However, while funding has been sustained and even increased, budget investments in social services have not entirely prevented their deterioration. Since the economic crisis there has been an increasing demand and need for social services especially for pregnant women, low income families, and truant youth. These challenges have led to an increase in collaboration across different sectors.

The Cuban government's recognition that collaboration is needed between government ministries, local levels of government, and community-based organisations signals an important change. Furthermore, the government has come to understand that despite its commitment to universal programs and accessibility, the current situation urgently requires decisive action, and it has thus implemented important community-based initiatives that focus resources on the most vulnerable sections of society.[23] This response represents a significant departure from past practices in service delivery and has gained support from the UNDP for its focus on human development at the local level. The UNDP initiative is designed to strengthen the local capacities of development (community, municipality and provincial), make local development environmentally sustainable, improve the coverage of social services for the population, improve gender equity, and support decentralisation (United Nations 2002: 3).

According to Uriarte (2002), critical problems with the delivery of services have emerged from lack of collaboration between the different sectors. In a highly centralised society like Cuba, this separation is less effective. Planning in every sector takes place nationally and directives flow down to the local levels of government in the provinces and municipalities. Typically, collaboration between

---

23   GDIC – Grupo para el Desarrollo Integral de la Capital.

the sectors is rare and this results in a kind of incoherence at the community level (Uriarte 2002). Because social programs operate concurrently within a community the lack of articulation of common objectives and methods has resulted in overlaps that often leads to the inefficient and ineffective delivery of social services.

The current move towards targeting the problems of vulnerable groups is a significant departure from the emphasis on 'universality' in the Cuban revolution. Providing for the needs of the poorest, most vulnerable families in Havana initially caused alarm to many Cuban residents. For four decades the prevailing revolutionary ethic had been that the government provided the same to all Cuban families – food with the ration card, universal medical care and education in the name of equity. Today, these changes have become more widely accepted as the means by which the government can implement preventative measures in order to avoid another externally applied Special Period or internally activated political dissent; the idea now is one of regulating these forces.

That Cuba is one of the last socialist strongholds in existence can be directly attributed to its unique form of socialism based on its geopolitical context. This model has developed in response to centuries of dependence on the large imperialistic nations of Spain, the United States and most recently the Soviet Union. The independence movement of the late 1800s, which eventually led to the 1959 revolution, was based on the concept of national sovereignty and José Martí's ethical ideal of social justice whose main objective was to build a republic 'with all and for the good of all' (Benjamin 1997: 113). Early revolutionary speeches by Fidel Castro and Che Guevara in the 1960s illustrate that the revolution was an attempt to build an alternative socialist model based on Cuba being a Latin American country, not a European one, and incorporating the ideals of José Martí. However, within the revolution there existed another school of socialist thought. Those educated in the Soviet Union conceived of a form of socialism based on the theoretical underpinnings of Leninism and Marxism as practised by the Soviet Union and the Eastern Bloc. As such, within the revolution there have existed different perceptions of socialism, with the last four decades of the revolution being a process of adaptation and transformation. From the outset, Fidel Castro argued that Cuban socialism would be an alternative to the colonial system, the injustices of Latin America, capitalism and the European model of socialism. As Benjamin (1997: 114) wrote:

> Only a humanistic and democratic definition would connect Cuban socialism to its own ideological tradition ... In the historical conditions prevailing in 1961 and in the face of US intransigence, socialism would be the only viable option to achieve independence with social justice.

Above all, socialism in Cuba, has been underpinned by ideas of social equity, and subsequent economic and social development deserve close scrutiny for its achievements. This is precisely what NGO study tours in Cuba set out to offer participants the opportunity to study.

The series of presentations, project visits and pre-tour information that comprise development-oriented tours are designed to provide comprehensive insight into contemporary Cuban realities, including the distinctive development paradigm with its mixture of successes and failures. As discussed in Chapter 2, the tours aim to provide participants with opportunities to meet local people and grassroots organisations and discuss local development issues, exchange ideas and establish meaningful relationships. The goal of the NGOs is to provide educational and culturally responsive tours and they begin this process by providing participants with in-depth literature prior to the tour. This literature is provided because the NGOs believe that it will lead to a well-informed tourist, which in turn will lead to fewer negative impacts typically associated with tourism in developing countries. To enhance this educational agenda, the tours provide a planned program to help participants understand Cuba from a development perspective in a way available to few other tourists. Global Exchange, in particular, aims to educate the US public on global issues and encourage active campaigning for international human rights, and for a change in US foreign policy towards Cuba. Another goal of the NGOs is to provide 'authentic' interactions with local communities. Participants come away from these interactions with deeper understandings of the struggles and accomplishments underpinning local development initiatives.

While the aim of each tour is to provide opportunities to meet with local people and grassroots organisations to discuss development issues, notably, the tours do not engage with local dissidents despite there being a burgeoning and increasingly public body of critique. Dissidence in Cuba has been fraught with inconsistent reactions by the government varying from hard line policies imposing lengthy jail terms on people opposed to Cuba's Marxist regime to a more encouraging trend in the late 1990s towards greater tolerance of protestors. Whilst I was living in Cuba in 2002 the former US President Jimmy Carter visited and met with dissidents. Further, in the past decade some of the well-known dissidents have even been allowed to travel abroad. While the Cuban government disapproved of the Varela Project, which calls for a referendum on greater political liberties and economic reforms, it did not imprison Oswaldo Payá who began the Project in 1998. This might be because he is internationally acknowledged as recipient of the European Union's Sakharov Prize for Freedom of Thought. In 2003 the Cuban government initiated a massive crackdown on dissidents, including an 18 year jail sentence for Oswaldo Alfonso Valdes, the leader of the democratic Labor Party in Cuba, a group supporting multi-party elections, free speech and free enterprise. The most contentious of recent arrests was that of 75 people accused of 'subversive activities' associated with the US Interests Section[24] in Havana. James Casson, chief of the US Interests Section confirmed "It's true, the United States is not a passive observer [of the growing opposition in Cuba]. After all, our goal is the rapid, peaceful transition to democracy. Yes, we gave them books, access to the internet, and a place to meet" (Overington 2003). Chillingly, in 2003 three men

---

24   Somewhat similar to an embassy.

who tried to hijack a ferry to America were executed by firing squad. Such events have caused great alarm among international human rights activists.

In 2002, Marta Beatriz Roque, another leading dissident who had served a three-year prison sentence, announced a new opposition council called the Assembly to Promote Civil Society, an association of over 300 groups, to co-ordinate opposition efforts. The Assembly has, however, attracted no public resistance from the government. In May 2005 the inaugural meeting of the Assembly drafted and disseminated a 10-point resolution for a transition to democracy. It laid out several benchmarks for liberalization, including economic reforms, genuine pluralism, abolition of the death penalty, and the release of all political prisoners. Interestingly, the government made no arrests of anyone over the two days of the meeting. However, Cuba's highest profile dissident Oswaldo Payá, leader of the Christian Liberation Movement, did not attend the Assembly claiming it was extremist in its goal to scrap the Communist system entirely in order to chart a new democratic framework. Payá's Varela Project and National Dialogue initiatives aim to promote democratic change within Cuba's post-1959 constitution. Another critical difference between the two is that while Payá opposes the US embargo, Roque and the Assembly support it enthusiastically. Such rifts highlight potential dilemmas for 'would-be' democrats. One thing that was made clear by many Cubans I interviewed is that few would be willing to give up all the benefits of socialism, such as free schools and medical care. Indeed, many Cubans are proud of Cuba with its focus on the ideals of sharing above all else and which is depicted as a direct antithesis of capitalist societies.

The Cuban government denies that it represses freedom of speech, or that it holds any prisoners of conscience, arguing that those imprisoned government opponents are there on legitimate charges including sometimes violent counter-revolutionary activities. Castro's usual response has been that many other countries, including some in Europe, have imprisoned people for illegal political activities. As the Cuban government has displayed rapid changes in its stance on dissident activities and because development agencies aim to remain apolitical the NGO study tours do not include visits with dissidents, nor embrace discussions of political dissent. This does not mean tourists are unaware of the perils of dissent. However, the primary focus is on the revolutionary achievements, or it could be suggested socialist ideals and reverential visitations.

Cuba's model of development, the US-imposed embargo and its effects, the Special Period, the subsequent economic reforms and social changes coalesce to provide a distinctive situation in this era of global interconnectedness. The nature of its revolutionary ideal social development and active participation by local people has been of interest to Western intellectuals and political pilgrims (Hollander 1981) disillusioned with Western society since the 1960s and 1970s. As Hollander notes, "the utopian susceptibilities of contemporary Western intellectuals are part of a long-standing tradition" (1981: 29). While the tourists involved in NGO study tours are not necessarily political pilgrims in search of utopia there are some very strong parallels between the early reverential tours to the Soviet Union to see

the ideals of socialism in action by those critiquing capitalist development and
NGO study tours to Cuba today. As a result, the following chapter explores some
excerpts from a development-oriented tour in order to provide a sense of what it
is like to participate in a NGO study tour in Cuba including what the tourists are
seeking and what versions of the revolution are being mandated as successful.

Chapter 4

Tourist Encounters with Endogenous Development

I arrived in Havana in the evening. It seemed to take hours to get through immigration and I remembered how intimidating the brusque immigration officials can be. Eventually I emerged into the night; the humidity was stifling. I took a taxi through the dark outskirts into Havana gaining my first views of the city. As we approached I noticed some familiar notable landmarks: *Plaza de la Revolución* where Fidel Castro had often addressed thousands of people; the grand stairs and Romanesque pillars of *La Universidad de la Habana*. I recalled that driving into Havana was different from driving into other international cities because of its distinct lack of advertising billboards. Instead, billboards espousing communist ideals and principles marked the drive into the city. One might expect to see images of Fidel Castro everywhere but instead busts of José Martí,[1] Cuban writer, poet and father of Cuban independence were abundant. I saw a billboard stating a famous Martí quote "*con todos y para el bien de todos*" – "with all and for the good of all". We drove along the seafront past spectacularly dilapidated buildings, with the ocean spraying across the road as waves crashed against the seawall known as *el Malecón*. I arrived at my *casa particular* to find Sofía sitting by the window awaiting my arrival. I was returning to Cuba after a year long absence to co-ordinate a 16 day NGO study tour for a small group of 12 Australian and British tourists to experience Cuba and its social development.

The significance of NGO study tours in Cuba is comparatively small in relation to mainstream tourism. But as a niche, it is an expanding area of tourism that has wider benefits for Cuba than just economic as I demonstrate throughout the remainder of the book. The intention of NGO study tours is to show the positive sustainable development initiatives being undertaken in Cuba by ordinary people and thus inspire tour participants to become more socially and environmentally active on their return to their own countries, while also providing some of the characteristics associated with conventional tourism. The tours included seminars associated with visits to community groups and NGOs, provided participants with an insight into daily life in Cuba, and they included significant recreation and sightseeing. At the core, they offered an insight into social development while discussing the merits, failures and challenges of Cuban socialism. The social

---

1 José Martí, claimed to be Cuba's founding father was well known as a writer and poet when he formed the Cuban Revolutionary Party, uniting various revolutionary factions in the late 1800s. Plaster busts of Martí are found all over Cuba, particularly outside schools.

development that the revolution strived for is becoming unique in this age of globalisation and neoliberalism because it offers very particular insights into alternative models. Cuba's political, social and economic environment is changing and its future is uncertain. The end of economic and military aid from the former Soviet Union led to a search for new money, new alliances, and new ways of doing things. This has allowed NGO study tours to take place in Cuba for the express purpose of looking into how people are addressing social development problems and what they think about their future.

This chapter offers details from selected portions of a NGO study tour. I intend it to operate on several levels by offering an insight into social development in Cuba, demonstrating in what ways tour participants learn about development and offering a sense of the specific sites visited. My intention is not to provide a critique of the presented material but a detailed description of the types of issues the tourists are exposed to in order to lead on to analysis of their experiences.

As noted earlier, these tours are based on a series of prescriptive meetings about social development and because they take place in Cuba, social development is framed within the context of socialism. Although the focus of NGO study tours is on development, the political context within which they take place deserves close attention as it sheds light on the motivations of the government in promoting this style of tourism. We can draw parallels between the political tours to the Soviet Union in the 1930s by Western intellectuals in search of utopia. Hollander (1981) refers to the "techniques of hospitality", measures intended to influence the perception of the international visitor of the country visited and which act as a control over their experiences. He explains that the political tour was one such technique designed to highlight to the international tourist "carefully selected sights, events, institutions, groups and individuals and isolating him from others not conducive to a favourable assessment of the social system he was being introduced to" (Hollander 1981: 372). Thus tour itineraries were tightly controlled with little free time.

The highly organised nature of political tours to the Soviet allowed for a selective presentation of reality, one which increased the likelihood of productive experiences. Political tours were purposefully designed to systematically expose tourists to those qualities of the country that would ensure a good impression.

> Even if the visitor harbours any abstract or generalised notions about the possibilities of social injustice, material scarcity, or institutional malfunctioning ... the visible, tangible realities he comes in contact with powerfully counteracts his apprehensions ... What the visitors are in no position to know is how typical or how characteristic such sights and impressions are, or how adequately they convey the flavour of life in the country at large. (Hollander 1981: 17-18)

Hollander criticises the intellectuals who went to the Soviet Union during the 1930s for not applying a critical eye. He says that tourists on political tours leave a country with potentially distorted views about healthcare, education and the like, based on only several visits to health clinics, schools and other institutions; in the Soviet case, "they rarely confronted the self-evident limitations of their experience and its implications for generalising about all they had not and could not have seen" (Hollander 1981: 20).

With this in mind, the contentious issue of human rights in Cuba emerges as an example of whether tourists on NGO study tours confront the limitations of their own experiences. Learning about the government's political stance on freedom of speech, dissidence, or political prisoners is not a central focus of the tours. As we shall see in later chapters, based on their experiences and the knowledge gained, tour participants seemingly reconcile Cuba's human rights issues with the visible social achievements they experience. For these tourists the government's position that the provision of free healthcare and education constitute the fulfilment of fundamental human rights and, perhaps, this provides justifiable grounds for visiting Cuba and exploring their political and social attributes.

The style of such tours and their intention is a domain worthy of critical scrutiny. The Cuban government clearly views international tourists as political agents who can give 'first-hand' accounts of the country and its achievements. In the same vein as the previous political tours to the Soviet Union, NGO study tours are a means to the development of international relations more receptive to Cuba's specific economic and social policies.

As mentioned, an important aspect of this style of tourism is the provision of a comprehensive selection of reading material prior to the tour, aimed at preparing tourists for their encounter with the country's political and social system. This is so that they can begin to gain a more informed understanding of Cuba and to assist in formulating questions that they might want to investigate while there. The readings consist of a collection of articles from varied sources including academics (both Cuban and international), newspapers, websites and NGOs. Once in Cuba, the tourists witness many things. How much they absorb is another thing, but they are alerted to the many particular and unique aspects of Cuba's recent history and development.

In particular, the tourists encounter different aspects of Cuban social development. For example, some of the tours visit the Cuban Federation of Women, a non-governmental organisation that has brought about major advances for women in Cuba over the last 40 years. Issues such as employment, equal rights, family, health, and social prevention (for example, sex work and domestic violence) are addressed in this meeting. At other times, the tour group has a free day in Havana, and while in the provinces, the group visits the Zapata Peninsula to learn about ecology and environmental issues, and the National Association for Small Farmers (ANAP) to learn about sustainable agriculture and some local co-operatives. They visit the Che Guevara museum in Santa Clara and have free time in Varadero, one of the largest tourist resorts in the Caribbean, to experience the enclavic style of tourism that is synonymous with Third World islands. They also meet with the NGO CubaSolar to learn about the extensive work on renewable energy being done in rural areas. Towards the end of the tour, the group travels to the western province of Pinar Del Rio via the Reserva Sierra del Rosario to visit the Las Terrazas Community Project where to learn about the reforestation of the region and the sustainable development of the community. On its return to Havana, the group meets with the Minister of Tourism to discuss their experiences and the contrasts they witnessed between Varadero and the provinces. It meets with the Reverend Raul Suarez of the Martin Luther King Centre to discuss Cuba's vision for the future. It is not possible here to detail the entire itinerary of all the tours, which comprise one to three weeks of development-oriented presentations and visits. However, the following selection of excerpts from different tours was chosen to highlight the community development aspect of the tours. I've also focused here on education and health visits because each tour had a significant number of participants who were professionals in these fields.

Usually on my first day in Havana I would attend to some final tour preparations. In the morning I would catch up with my Cuban family, and then visit the Cuban tour agency hosting the tour, *Amistur*, following which I would visit Rose at the Oxfam Canada office for a *cafecito*. In the heat of the afternoon I would sink into the languid rhythm of Cuba while waiting at the bank in one of Cuba's ubiquitous queues. This bank was located on the ground floor of the famous Havana Hilton (now called *Habana Libre*), where Fidel Castro and Che Guevara first set up office at the outset of the revolution. All foreign visitors seemed to exchange their money there, so the queue was long and seemingly stationery. I made a mental note to find a bank away from the tourist areas where the queues might be smaller.

Day Two, Morning – Presentation and Discussion:
Social Development in Cuba

The first full day of the itinerary would be busy with presentations by three key solidarity and development institutions. Most of the tour participants would have flown in the previous day and joined my evening orientation meeting with our Cuban tour guide, Emilio. On this occasion, the participants were staying in the old part of Havana in a small boutique hotel. The rooms had balconies with long vines draping into the internal courtyard. After a breakfast of *tortilla con queso y jamon* (cheese and ham omelette) and strong sweet Cuban coffee, we piled into a mini bus and headed off for Vedado, a previously affluent neighbourhood of Havana where I lived with a Cuban family. Some members of the group could not help but notice that most of the houses in this quarter, were in a good state of repair compared with the dilapidated buildings on *el Malecón* near their hotel. We went to Vedado to visit the headquarters of the Cuban NGO 'Institute for Friendship between People' (ICAP), the organisation which hosts study tours in Cuba. ICAP was housed in a magnificent old Spanish mansion. We were welcomed by Alicia Corredera, the Director of Asia Pacific Division and a strikingly elegant woman, who escorted us into a palatial room of white marble floors, intricate frescoes and an impressive crystal chandelier; a clear indication of the wealth in pre-revolutionary Cuba. We were seated around a grand polished antique table for our first foray into Cuban solidarity.

Alicia, speaking fluent English, gave us a presentation about the role of ICAP and what its role with our tour group was. First, she explained that most mansions, such as the one housing ICAP, had been converted into places of business, schools, and foreign embassies. ICAP is a social organisation founded in 1960 to strengthen the bonds of friendship and solidarity between the people of the world. From the beginning, she continued, ICAP helped co-ordinate the International Brigades which came from around the world to support Cuba by offering aid in agriculture, construction and other sectors of the economy. Nowadays it continues to facilitate visits and exchanges between friendship organisations and NGOs. By hosting tours and delegations, Alicia explained, ICAP disseminates information about Cuba in order to correct any misconceptions; in this sense, tour participants and delegates become Cuba's ambassadors. The Institute organises delegations for international groups, similar to ours, to visit Cuba interested in learning about the Cuban context and the social changes taking place in the country. It promotes specialised visits to Cuba through its travel agency, Amistur. Groups such as ours, we were told, are mainly interested in the sociopolitical aspects of Cuban reality. The *Casa de la Amistad* (Friendship House), run by ICAP, organises social and cultural events and exchanges. ICAP also receives and distributes humanitarian aid from international solidarity groups, and from tour groups paying a small donation to each of the organisations that they visit, in gratitude for taking the time to talk with international visitors. Most people in our group had brought some form of aid with them. We were advised that we could either leave our aid with ICAP, who would disseminate it on our behalf, or, we could distribute it to the schools, hospitals and community projects that we visited. Some people chose to give their donations to ICAP, while others preferred to wait until we travelled into the provinces to distribute their gifts personally to schools, hospitals and community projects.

After our reception with ICAP, we visited another NGO, the *Grupo para el Desarrollo Integral de la Capital* (Group for the Integrated Development of the Capital – GDIC). Our tour guide Emilio informed our tourists that this was a good starting point for our visit in Cuba because there was a large scale model of the city (*la maqueta*) housed there which would help to give us an overview of the city. GDIC is a non-governmental civil service institution founded in 1988 that addresses issues of urbanism and works to preserve the architectural patrimony of the city, improve living conditions, assist in the development of local economies, provide urban education and develop community pride and identity. Essentially their work centres on the economic, cultural and social development of the capital, Havana.

Surprisingly for a centralised State, there are a substantial number of Cuban NGOs, but their role is not typical of NGOs that usually fill a void left by governments. Rather, NGOs in Cuba (supported largely by international NGOS such as Oxfam Canada works closely with GDIC) work with the government in community development in a time of need. It is worth noting that confrontational NGO's are not likely to operate in Cuba because it is a country that has carefully selected its affiliations with countries and organisations that are politically sympathetic to its cause.

A well known architect, urban planner and sociologist working with GDIC at the time, Miguel Coyula discusses urban development and infrastructure needs of Havana using the model. It is also an efficient way to quickly get an orientation of the city. *La Maqueta de la Habana*, Miguel pointed out, was constructed from the recycled wood of old cigar boxes. With a surface area of 144 square meters representing 144 square kilometres, it was at the time, the third largest of its kind in the world. The model includes every building in Havana and even today serves as a reference tool for commercial and State urban developers. GDIC thus plays an integral part in advising on new construction and urban projects.

Once we had made our way around the model, we sat down with Miguel who discussed the emergence of the neighbourhood movement – community organisations based in Cuban *barrios* (neighbourhoods) created to address urgent needs of those enduring the hardship of the economic crisis. Miguel informed us that, throughout the revolution, the Cuban government structure has been highly centralised. It focused resources on national priorities, which often resulted in local problems being overlooked (c.f. Uriarte 2002) and unresolved due to complexities that do not exist at the national level. Miguel explained that we would no doubt notice this in numerous urban areas of Havana during our tour. Policy development and assigning resources has been the domain of the Cuban government, which relies on the expertise of think tanks and the National Assembly in order to make the most informed decisions on public policy for the entire country. However it is different from Western top-down models in that the government is responsible for social services and works with grassroots organisations that also have international assistance to address local problems.

Various national organisations operating at the local level provide opportunities for Cuban citizens to have an effect on policy development, but their effectiveness is limited. The purpose of these mass organisations is twofold, to convey information about local issues to the government and to deliver the directions of the initiatives to be undertaken in order to rectify local problems. Local branches of these national organisations are located in every neighbourhood and are specifically aimed at addressing issues pertaining to a broad cross section of society including women (Federation of Cuban Women – FMC), neighbours (Committees for the Defence of the Revolution – CDR), farmers (National Association of Small Farmers – ANAP) and youth (Pioneers, Federation of Middle School Students, and the Communist Youth). These organisations enable local people to discuss problems within their neighbourhoods regarding social services, education and healthcare. Miguel pointed out that one of the main goals of the tour upon which we were about to embark would be to meet with such groups in Havana and the provinces.

Although community development has existed in the form of mass community organisations in Cuba for a long time (as is typical of socialist countries), it has taken on a different form since the 1990s. The main community organisations in Cuba offering a wide variety of services have been typically located in every neighbourhood block. The FMC, for example, co-ordinates comprehensive vaccination programs and oversees public education in local communities. CDRs have been established to provide primary support at the neighbourhood level, taking responsibility for the security of the neighbourhoods and organising volunteer work and community activities. CDRs ensure the fulfilment of communitarian tasks such as recycling, night patrol of the street and some health campaigns. The CDRista, the person nominated by the neighbourhood to head such activities, has a ledger listing everyone on the block and keeps an eye out for 'antisocial' behaviour. One of the tourists immediately inquired "What happens to someone who is antisocial?", Miguel replies that they are reported to police who gives the person a warning and keeps a watch on them. The person could then be liable for up to four years of sanctions if they are in any way antisocial. It became immediately apparent that the CDR is as much an annex of the police as an organisation of the community, benevolent neighbourhood organisation or grassroots surveillance of fellow citizens? The family doctor and nurse are another fundamental initiative at the neighbourhood level, providing primary health care for all local residents. The local schools are in use throughout the entire day with many local school children attending after-class activities organised on the school grounds.

Miguel explained that it was these community organisations at the block level that provided the fundamental social support needed during the economic crisis of the 1990s. When times were particularly tough for residents, the CDRs were active in providing recreational activities at the block level, organising neighbourhood cleanups, and instituting a neighbourhood watch to aid in the prevention of crime and delinquency in the neighbourhoods. Likewise, at this time, FMC distributed vitamins to every household and organised community meetings to discuss the ongoing issue of food shortages and to provide forums for discussing different ways to cook meals with the limited food available (Uriarte 2002). This kind of community assistance also provided a sense of moral support when times were particularly difficult.

Although mass organisations were a long established and important element of the fabric of Cuban society, during the Special Period they were evidently not the best vehicles for creating solutions to the social problems because they could only act on *orientaciones* (directions) from the central government. As a result, they were unable to provide leadership in addressing the community problems (Uriarte 2002). Long-standing community problems were exacerbated by the emergence of new difficulties, such as the lack of resources. Thus the nature and scale of social problems increasingly extended beyond what the mass community organisations could address.

The local level in Cuba previously referred to the municipalities (cities, towns, villages within the provinces); however, the economic crisis highlighted the fact that the municipality was far too large and the needs of the population far too diverse to be addressed by generic solutions proposed at that level. Social needs became more obvious as resources during the Special Period declined, leading to great difficulties in responding appropriately to the many demands of the community. This vacuum at the community level fuelled the emergence of a neighbourhood movement consisting of community organisations based in Cuban *barrios* (neighbourhoods) to address the urgent needs of those enduring hardships as a result of the crisis. The neighbourhood movement changed the focus of social urban development from a purely vertical model to one incorporating horizontal networks at the community level thereby integrating the government, the mass community organisations, Cuban non-government organisations like the Martin Luther King Centre and the Group for the Integral Development of the Capital (GDIC), international development organisations, academic institutions and most fundamentally, the residents. Miguel explained that the *Tallers de Transformación Integral del Barrio* (Neighbourhood Transformation Workshops) and the *Consejos Populares* (Popular Councils) were the first organisations to engage in this urban development work during the Special Period.

GDIC established a series of neighbourhood transformation workshops, *Tallers para la Transformación Integral de los Barrios*, in three neighbourhoods to promote the development of urban agriculture, social prevention work and conservation and to provide specific action guidelines for various community groups with broad participation in decision-making. These three neighbourhoods, Atarés, Cayo Hueso, and La Guinera (Miguel indicated their locations on *la maqueta*) experienced considerable social and physical deterioration. The workshops (of which twenty have formed over the last two decades) were developed using the following procedures: to conduct a preliminary study to identify neighbourhood problems; call for a community meeting with volunteer participation to explain the workshop's aims and purposes, its characteristics and what actions may be undertaken; explain to residents that changes will only occur with their full consent and active participation; incorporate and seek the commitment of residents in the solutions of their own community problems and supplying residents with tools and professional advice to become actors of change within their own living environment. Miguel informed the group that they would visit a *Taller para la Transformación* in the community of Atarés as one of their project visits in the coming days. This visit had been incorporated into the itinerary as part of the tour's goals of facilitating people-to-people encounters and inspiring tour participants to become more socially and environmentally active.

Each workshop enlists the advice of professionals to address the problems of the neighbourhood. One of the primary objectives is to identify the most significant problem areas within the neighbourhood and to mobilise resources to address them. Housing is often the main problem and workshops make it a key goal to address this. It has become the role of the workshops to identify the most needy tenement buildings and find alternative temporary housing for the residents during renovation.

This initiative is most effective in bringing resources into the neighbourhood with government support, providing the workshops considerable leverage in obtaining resources to address identified problems at the neighbourhood level with significant input by community residents. Gradually this neighbourhood transformation initiative has expanded into other vulnerable neighbourhoods in Havana. The Special Period had a major impact on this enterprise as construction materials became limited and the infrastructure development approach was transformed into interventions focusing on the social needs of the neighbourhood residents.

In order to carry out their new role, workshops applied participatory community planning methods which allowed needs assessments to be conducted. Residents were involved in the process. With the help of professionals, the focus was shifted to the most vulnerable members of the neighbourhood, reinforcing the cultural identity of the neighbourhoods and working to improve the neighbourhoods as much as possible with scarce resources. Work conducted included the renovation of resident housing and schools, the construction of doctors' offices in each neighbourhood, environmental projects, economic development workshops for neighbourhood artisans, job training and computer classes for women, social service activities such as mothers groups, self esteem groups for women and after school programs focused on local cultural expression, and so forth. The workshops' ability to identify local problems enabled other mass community organisations to become more efficient in their roles, rather than replicating the efforts of the CDR and FMC and the health system.

The Special Period saw the flourishing of social development experimentation through community participation, community development, and the use of participatory community planning methodologies. The neighbourhood transformation workshops worked closely with the Popular Councils within the neighbourhood movement. The Popular Councils were formed at the height of the economic crisis to help fill the gap between the neighbourhoods and municipalities by focusing on horizontal networks at the local level. They are composed of volunteers elected by the residents of neighbourhoods as well as by representatives of the main economic, social and service institutions including the mass community organisations. Their job is to monitor government activities for the purposes of improving services at the local level. As the central, provincial, and municipal governments became ineffective in addressing local problems during the Special Period, the Popular Councils took on the role of overseeing and managing the impact of the crisis at the local level. Some entrepreneurial Popular Councils have instituted community development projects focused on improving people's nutrition through urban agriculture, environmental awareness campaigns, cultural programs, and so on.

As mentioned earlier, a central aim of the tours is to provide participants with opportunities to meet local people and grassroots organisations to discuss local community development issues. Tour participants experience Cuba from a human development perspective in a way available to few others. During our tour, Miguel provided comprehensive information about community development, which gave the tour group the necessary background for visiting community projects such as Atarés, El Comodoro and Pogoloti (all supported by Oxfam). Community development initiatives in Cuba have some key characteristics. Firstly, initiatives operate in small geographic areas targeting specific communities, unlike the large-scale directives that target large regions. Typically, strategic community planning begins with a participatory diagnosis of the major issues facing the local residents of a neighbourhood conducted by the workshops, Council and residents. Once this diagnosis is completed, a series of meetings are organised where the Council and community based organisations analyse the findings and determine the strengths, weaknesses, opportunities and threats they face in addressing the problems. This technique assists the group in prioritising those areas in which they have the potential for success. Miguel advised the group that we would learn more about community diagnosis when we visit some transformation workshops.

Secondly, like so much of contemporary development around the world, an important characteristic of the design and implementation of projects is participatory community planning. The GDIC, Ministry of Culture, and Martin Luther King Centre all support the Popular Councils in applying these methodologies in the Cuban context. In Cuba, there is a long history of political participation and social voluntarism, such as the Literacy Campaign and the vaccination of children, which were administered voluntarily by ordinary citizens. Previously opportunities for residents to participate in the decision-making process had been extremely limited because initiatives had been centrally planned. Participation by residents who are impacted on by the projects has become a new core element of community development processes in Cuba. Although most Cuban development projects strive to incorporate local participation, in many cases, decision making power has not been fully realised by residents.

The third and most important characteristic of community development in Cuba is the undertaking to prioritise neighbourhood resources for the use of community development initiatives supported by both small government and/or international NGOs. International NGOs like Oxfam are supporting this initiative, which enables tourists to participate in development-focused tourism in Cuba through a program of visits to these community projects. By visiting community workshops tourists support the prioritisation of neighbourhood initiatives by exchanging information with local people and by additional donating resources to support their activities. In order to obtain other resources, projects have to rely on mass community organisations such as the CDR and FMC and on government agencies like the Ministry of Culture or the Ministry of Health. The Ministry of Culture, for example, funds the *Casas de la Cultura* (Houses of Culture), which in turn finances local artisans and provides performance space. Community projects are also supported by local universities, research centres, Cuban NGOs, the Council of Churches, and the *Centro de Información y Estudio sobre los Relaciones Interamericanas* (CIERI – Centre for Information and Study of InterAmerican Relations). International NGOs also fund community development projects.

It was not until the mid 1990s that international NGOs began to operate in Cuba, with their work directly supporting the work of national NGOs rather than State institutions. Typically the emergence of NGOs in Latin America, the Caribbean and elsewhere is due to structural adjustment and privatisation leading to the deterioration of social services (Benjamin 1997). In the vacuum created by the ensuing lack of resources, private voluntary agencies have stepped in and sought foreign funding assistance. However, the NGO movement in Cuba has followed a different course. As hinted at earlier, NGOs do not operate independently of the State in the delivery of services; instead they are actors in the government-sanctioned development process, engaging both the community and the government.

The government did not initially embrace either Cuban or international NGOs. This was because, until the 1990s, Cuba did not need assistance because its favourable terms of trade with the Soviet Union and the Eastern Bloc had helped to finance investment in social services that were unparalleled in developing nations. In addition, the government wanted to enforce political control in an attempt to prevent outside interference. Cuba considered itself an aid donor, not a recipient country[2]. International co-operation was not particularly welcome because it had been characterised in Cuba by political conditioning, paternalism and cultural imposition (Sinclair 2000).

Since the 1990s, however, the work of NGOs has been fundamental to the success of community development projects throughout the economic crisis and has emerged as an incipient grounded movement that puts into practice the values upon which the revolution was founded. As Miguel explained to the group, Cuban NGOs, supported by international NGOs, engage the community in programs based on social justice, compassion, solidarity and participation. He informed us that this was exactly what we would see during our visits to various community projects. The purpose of such immersion in community projects, he explained is to encourage the exchange of development ideas and knowledge with the locals. Ideological contamination is not considered to be a risk because such tourists supposedly go to Cuba with either a neutral attitude or favourable predisposition (c.f. Hollander 1981).

Finally, Cuban community development initiatives are characterised by the support and push for local grassroots leadership and efforts to build capacity at the local level. Initially people learned through experimenting with the support of national NGOs and community mass organisations; later national organisations began to synthesise the early lessons and with support from international NGOs began to teach more people their learnings. The capacity developed at the local level is a direct result of the process of community development, and indirectly tends to the strengthening of other aspects of the neighbourhood as newly learnt skills are transferred to other community organisations.

---

2   This is still a point of contention for many Cubans who feel that the government's financial support of other countries should be directed at local Cuban development.

Members of the group responded to Miguel's in-depth discussion with a range of questions. There was particular interest in the topic of the provision of housing in a socialist context. Can people buy and sell houses? Do Cuban people own their own homes or does the government? Has housing been one of the biggest problems for the revolution? How is the Cuban government addressing the housing crisis? Miguel smiled a knowing smile, obviously having often answered such questions from foreigners. His responses can be summed up as follows.

One of the greatest and ongoing challenges for the Cuban revolution has indeed been the adequate provision of housing for its expanding population. Presently there is a housing shortage, and in many instances existing housing is extremely poor. The government has reduced the cost of housing and has initiated programs for building more homes. Housing in Cuba is not an entitlement, but rather a benefit. Consequently, many families will have multiple generations living in the one home. People cannot legally buy and sell a house or an apartment privately in the market. If they want to move, they go down to the crowd on the median strip on Paseo del Prado in Habana Vieja to read the swap notices nailed to trees. But swapping one's house is not simple. A change of residence permit must be applied for and then an inspector must verify that the number of people living in a house meets the rule of ten square meters per person. Residents wanting to move must register with the *Committee for the Defence of the Revolution*, have their identity card changed accordingly, and go to the ration book office and the driver's license office to obtain their new documents.

At the outset of the revolution, the issue of housing was high on the agenda. Rent reductions for tenants were instituted. The government has capped rents at 10 per cent of a typical family income, but those who are sick or elderly pay even less. According to Miguel, currently less than 10 per cent of Cuban families rent as most Cubans now own their home. The Minister President of the Central Bank of Cuba has said that 85 per cent of the population have the ownership deeds of their homes and the remaining 15 per cent are given advances of up to 20 years with an interest rate of between 2 and 3 per cent to pay off their mortgages (Oramas 2001). This announcement, however, conflicts with the fact that many Cuban households hold three or more generations because younger Cubans cannot obtain their own house due to the severe shortages.

The housing shortage remains a serious issue for the government and this can especially be seen in many areas of Havana. Miguel explained that since the revolution, it has been the responsibility of the government to build and maintain housing. A significant amount of housing stock was built across the country in the 1970s and 1980s by construction micro-brigades. These micro-brigades were delegated by the government and consisted of a combination of professional and volunteer construction workers who volunteered to build very simple housing that was then distributed to the families of the workers. Currently, municipalities supervise the micro-brigades. Gradually, families have become much more proactive in constructing and renovating their own abodes.

A community development project in Cayo Hueso stands out as one of the country's most successful housing projects. Centro Habana, located in central Havana, is the most densely populated municipality in Cuba and has the largest concentration of tenements, or *ciudadelas*[3] in Cuba. As a result of such poor housing, the neighbourhood has extremely inadequate sanitation and unsafe drinking water, and consequently suffers from prevailing communicable diseases.[4] With the support of GDIC and the International Development Research Centre, a Canadian research institute, the neighbourhood of Cayo Hueso commenced a community intervention program aimed at improving housing, water supplies and the development of community centres (Campbell 2001).

The tourists were particularly interested in the topic of housing because of the stark contrast they observed between the housing on *el Malecón,* which was extremely dilapidated and the housing around GDIC, which is located in what was once one of the wealthiest neighbourhoods in Havana, Miramar. Here, grand mansions surrounded by landscaped gardens overlook wide boulevards. We lunched at a restaurant called *el Palenque* – the name given to communities of escaped slaves who established themselves in the Cuban countryside. The food here was an array of taro, potatoes, white rice and black beans (also called Moors and Christians *moros y cristianos*), fresh vegetables, fruit, pasta, roast pork, fried chicken and fish. We sat outside under straw roofs surrounded by tropical palms and I noticed there were many Cubans eating here too. Over lunch many of the members of the group engaged in discussion about the strength of social capital in Cuba in the form of mass organisations and the enormous voluntary campaigns such as the vaccination programs. They discussed the social relationships that these organisations foster and reflected on the nature of neighbourliness so central to these social networks in which Cuban people are involved, such as FMC and CDR. They mused whether such social cohesion would be possible in Australia or America. Certainly the message presented to us was not that mass community organisations are an integral aspect of oppressive totalitarian societies, but rather that they represent strong social cohesion which supports collective values rather than individualistic ones.

---

3   Ciudadelas date back to the last century as housing for the very poor made up of a series of rooms in which several families live. They share water and a common bathroom, which are usually located outside.

4   IDRC – country profile/Cuba 2003.

Figure 4 Dilapidated housing on *el Malecón*

Figure 5 A seminar during a tour of a FMC-run health project

**Figure 6 Elderly women at a community centre performing
     for the tour participants**

**Figure 7 *Las Terrazas*, a sustainable community project in
     Pinar del Rio province**

Figure 8 A formerly elite golf club now the Higher Institute of Arts
**   that we visit**

Figure 9 Visit to *El Comodoro*, a community housing project
**   for displaced families**

Figure 10 Outside the community house at *El Comodoro*

Figure 11 Tenements in La Habana Vieja

Figure 12 Tenements in Centro Habana

**Figure 13 Children playing between the dilapidated buildings
            along *el Malecón***

Day Two, Afternoon – Presentation and Discussion: Oxfam in Cuba

After lunch we met with Rose from Oxfam Canada. Some of the group were intrigued to hear that she has lived in Cuba since 1990 and was raising her two children in Havana. Rose addressed the role of Oxfam in Cuba and the unique aspects of Cuban social development. She explained that the Oxfam program includes some 35 projects and partner relationships with at least 12 Cuban organisations. Currently, seven countries within Oxfam International are participating in the program in Cuba, namely, Canada, Belgium, US, Germany, Spain, Holland and Great Britain. Rose explained that "Oxfam likes to work in Cuba because the island is considered in some way a school or laboratory for those seeking alternatives to the capitalist economy". Cuba, she describes, "is a very special place with a very special dynamic" – a poor country undergoing development, where feeding the population is still a struggle, and an essentially socialist system trying to adapt to a globalised, largely capitalist, world. There are many changes occurring in Cuba, which for Oxfam has more relevance as a model, than other Latin American countries that underwent structural adjustment by the International Monetary Fund and World Bank. For these reasons, Oxfam places great importance on making a positive contribution to development in Cuba to learn from it and to share this knowledge with other countries where Oxfam works. For Rose, there are many reasons from a development point of view to work in Cuba. For example, the resources Oxfam can supply are put to very good use in Cuba. "Sometimes there are difficulties in getting projects started, but Cuban people", she shared, "have a real appreciation for even small amounts of funding and they demonstrate that they can accomplish a lot with minimal funding, particularly because there is much talent and capacity here".

Rose informed us that Oxfam works in the areas of participation, economic alternatives, gender, disaster response, food security and human development, the last two being its main focus. Through food security programs, Oxfam promotes organic agriculture, small farm co-operatives, new initiatives, new technologies, women in positions of leadership in the co-operatives and general food security. An example of human development, she told us, is the work with the GDIC in Havana, but Oxfam also works with other organisations in this capacity. The majority of the human development work involves programs for training and small projects focused around issues of education and self-development. Oxfam's purpose in helping community development in Cuba is to support counterparts who have the will to change people's attitudes, beliefs, and behaviours. Specifically in the case of Cuba, Oxfam's priority for community development centres on sustainability through participation. It encourages people to take a more active role in the social change because in a country where the State has been very much a benefactor, planning has been centralised and now that is being transformed under the pressures of globalisation. Accordingly a lot of the work, we're told, in community development involves empowering people at the grassroots and organising processes by which people define their priorities and create solutions to their problems. Oxfam also supports the Cuban women's organisation, community-based organisations, study centres, think tanks, parks, universities and municipal and national government departments.

**Figure 14 *Organopónico* provides food for tenants in the soviet-style flats
         in the background**

After a brief presentation, Rose invited questions from the group. There was no shortage of questions and interest. An older woman Lily introduced herself and expressed an interest in understanding how social security works in Cuba. Rose described how Cuba's social security entitlement consists of State backed old age and disability pensions and survivors insurance for all workers. Pensions are proportional to salary and time worked. Prior to the revolution the system varied greatly in terms of benefits and relied almost exclusively on workers' contributions. Since the revolution, the program has been funded entirely by the government and the coverage is universal. Social security is a very real concern for a country whose population is ageing rapidly.[5] A longer life expectancy and the over representation of the young among emigrants are contributing factors. The World Bank indicates that, by 2040, those aged 65 years and over will account for 26 per cent of Cuba's population.[6]

---

5 World Bank Indicators, 2001.

6 Fransisco Soberon, Minister President of the Central Bank of Cuba detailed the average monthly costs for a Cuban family of four with two children under age seven at

The government provides social assistance through cash subsidies and special services to families in need of economic support. Typically, single mothers with young children in need of assistance for child care, families of deceased workers who are eligible for death benefits, and elderly workers who have not met the minimal time required for a social security pension are the recipients of social assistance. Special schools, services for the elderly, meals and laundry services for poor elderly and job training for mothers without a source of income are generally provided by the municipal entities (Uriarte 2002). In addition, Cuban people have access to a range of comprehensive universal subsidies provided by the government.[7] Subsidies cover the cost of water, gas, and electricity, food on a ration card, meals at schools and work places, day care for the children of working mothers and rents and mortgages. These subsidies add significant strength to the Cuban 'safety net'.

A gentleman who was sitting behind me, Tom, asked Rose to "explain how the ration system works?" There was a general murmur of approval for this question as people were curious about the ration card. *La libreta*, the Cuban ration card, was instituted to ensure that all Cubans have access to a basic food allotment at highly subsidised prices. International media often speculates that the ration card is evidence of how desperate Cuba is since the economic crisis of the 1990s. However, Rose explained that rationing had been a way of life in Cuba for 35 years prior to the crisis. *La libreta* lists all members of the household and what they have been provided each month, including food, toiletries, cooking fuel and in the past, clothes and shoes. The difficult economic conditions of the 1990s decreased the availability of food on *la libreta*, but the government maintained the supply of essential food items to every Cuban.

Some of the most basic commodities are heavily subsidised and available to Cubans through government stores called *bodegas*. Most food must be purchased in outdoors markets and supermarkets, mixing State and private sales. The markets accept Cuban pesos or US dollars while supermarkets (publicly owned) only accepted dollars. When shopping at a market you take your own plastic bags or you buy them from vendors at 4 for 1 peso. Meat (mostly pork) is displayed at markets unrefrigerated and uncovered. Once a month, there are massive outdoor sales of fresh food from trucks that bring the food in from the countryside. Prices are half of those in the regular markets and people will put up with very long lines to stock up. As an aside, Rose warned us that sugar is extremely popular, with massive amounts used in tiny cups of coffee and added to fresh fruit juices. Many workplaces provide food for their workers at heavily subsidised prices. Apparently

a total cost of 45.56 pesos ($2.07USD). This includes a monthly electricity bill of 13.60 pesos, telephone service of 6.25 pesos ($0.28USD), cooking gas 7.63 pesos ($0.35USD), and water 1.30 pesos per month per person ($0.6USD). The monthly rations for all Cubans includes rice, beans, eggs, bread, potatoes and powdered milk for children up to seven years (Oramas 2001: 3).

7   UNDP Human Development Report 2007/2008.

the food was better when the Soviet Union still existed because the alliance provided Cuba with a steady supply of much needed commodities at stable prices. As a result, the government was able to limit to an extent the effects of fluctuations in the capitalist world market making it possible to subsidise the cost of food and provide cheaper sustenance for people up until the collapse of the Soviet Union.

Day Four – Project Visit: Community Workshop

Arriving in Atarés to visit a *Taller para la Transformación del Barrio*, our first impression was how much this neighbourhood stood in stark contrast to the wide boulevards lined with sculptured trees and grand mansions of Miramar, where we met with Miguel at *la maqueta*. There were no lush gardens here, and the dilapidated grey buildings tightly packed together had not seen a coat of paint in many years. Stray dogs scampered through hot streets or lay listlessly on equally hot pavements. Local families in this neighbourhood lived in small overcrowded tenements with poor facilities. There were no symbols of the bygone wealth of other Havana neighbourhoods, it was clear that Atarés had always been a poor neighbourhood. We entered a building and were greeted warmly by a middle-aged woman named Lisa. The room was large and dimly lit with high ceilings of crumbling plaster. We sat on chairs in a circle and Lisa explained to us about the community workshop (*taller*). This visit had been arranged to showcase what Miguel had talked about on our first day, namely, community development and local participation through neighbourhood workshops, which accorded with the tour's overall goal of providing educational tours with a development focus. The project had commenced in 1988 thanks to GDIC's work towards the transformation of the city. The members of this *taller* had trained constantly to allow different methods to be applied and as strategic planning for the community.

Lisa told us that part of the research work involves trying to identify the needs of the community. These needs come from both the desires of the neighbours and the knowledge of the experts and the process of 'community diagnosis' is aimed at helping local people create a plan of action. Once those needs are established, different workshops are created to address various interests for sections of the community. For example, the self esteem workshops for women are aimed at helping women with daily life in a society that is traditionally dominated by *machismo*. They focus on bringing women together to discuss their issues and assist one another in finding solutions to their problems. There are also workshops for young children and elderly people focusing on literature or arts, enabling children to act and perform. Children are always in school from 8am to 4pm; schooling is free and mandatory and the workshops with children are co-ordinated with the local schools. There are always activities available. Lisa explained that Atarés has kept its traditions relating to religion of African origin with songs, dancing, cooking, and activities, all of which are maintained in the daily life of this community. The workshop is designed around this heritage.

A team of seven specialists, including architects, psychologists, sociologists, specialists in Afro Cuban culture, a specialist in accounting, all of them women, provides advice for this work. There is also a research social advisor and social worker. Lisa explained that the women in Cuba have proved to be more active than the men at the grassroots level. This she attributes to the community programs co-ordinated by the Federation of Cuban Women. Apparently, according to Lisa, this project has had some success but not as much as anticipated due to the economic crisis. The *taller* continues to call for the support of its neighbours and the government. This project has received much direct support from young people since 1999. Lisa explained that they have been managing the house we were sitting in and have greatly contributed to its success. Numerous groups have supported this project including many NGOs, particularly from Norway. Since the year 2000, the project at Atarés has relied heavily on Oxfam Canada to facilitate this work. Oxfam Canada provided finance for the workshop on literature, which allows groups to do readings at the house, and for buying chairs for small children. Also at this house there are activities aimed at developing the skills of children between aged two to five years. The project is led by a specialist from the Ministry of Education and aims to contribute to the social development of children. Oxfam Canada also supports the workshops aimed at keeping local traditions alive. Lisa explained that, in the neighbourhood, there is a certain feeling of belonging through the practice of traditions. The financing has enabled the *taller* to buy instruments and outfits for the musicians, and has also allowed the neighbourhood to build apartments and houses, and to repair others. In this way, the *taller* supports the efforts of the State.

The specialists who work here are employed full-time by the government. Miguel had told the group that there were now 20 of these transformation workshops in Havana that had operated over many years. This particular project has a very high level of participation by the local population in the identification of needs, the decision-making process, and the carrying out of tasks. At that time, the *taller* had recently helped in replacing sewage and water pipelines, which were constructed so long ago that they were in a severe state of disrepair. The workshop also works to repair the pavements, but Lisa informed us that there are still huge advances to be made in improving hygiene and preventing people from throwing their household rubbish into the streets.

Many questions arose from this visit. One tour participant wanted to know more about what exactly was involved in the literature workshop. Lisa explained that the workshop is aimed at developing reading skills among children. "This workshop is like a backup project to the Ministry of Education. During the holidays, the local children can come here to do both educational and social activities". Another participant asked "Can you explain a little about the elderly peoples' workshop"? Lisa explained that this is a joint workshop with the Ministry of Health involving a team of specialists who provide help to an elderly person.

> We provide logistical support. For example, they use this space for activities
> such as watching films, dancing and so on. Unfortunately there is not a very
> high rate of participation from the elderly. Since they only provide logistical
> support, there are other spaces for those elderly people who live too far for their
> old legs to walk.

One gentleman asked Lisa what the role of the architect is in the project. Lisa
answered him that the *taller* is actually connected to the urban development of the
community as well as to other organisations related to development, such as the
GDIC.

> The architect provides community work; actually she is one of the founders
> of this project. Every project has to be co-ordinated with what we have called
> a 'success communicator'. For example, when we talk about the workshop of
> self esteem for women we have to measure up the success of the activities of
> the workshop.

It was clear that the group was really interested in what they had seen. Afterwards,
Lisa invited us to look around the house and the various resources. It was at this
point that some of the delegates brought out their donations of stationary, pencils,
scissors, colouring-in books and such for the children's workshop. One lady
had a large bag with bundles of wool and knitting needles to offer to the elderly
ladies who come to the community house. The gifts were received graciously by
Lisa, who assured us that they would be greatly appreciated by the community
residents. As I wandered around the *taller*, I overheard some of the Australian
tourists chatting with Lisa and our tour guide about how this workshop could be
utilised as a model in poor socioeconomic urban areas of Australia, particularly in
the large housing estates subsidised by the government.

We clambered back into our mini bus and were on our way to visit another
community project, this one for people displaced by natural disasters. This
project is called *El Comodoro* and is located just on the outskirts of Havana. As
mentioned earlier, the housing shortage has been the central problem that the
Cuban government has struggled to overcome since the revolution. On the drive
to *El Comodoro*, the tour leader informed the tourists that this project emerged
from a need to house people whose homes had literally crumbled due to a lack
of resources for maintaining old buildings or as a result of one of the annual
hurricanes that batter the island.

Upon our arrival, we were met by a group of residents who escorted us into a
small white washed cement house with a garden. This was a 'community house'.
We took seats in the main *sala* and one of the ladies, Julia welcomed us. She told
us that this area had been created thanks to a social community project and that
the group of people meeting with us was a small representation of the people
involved in the establishment and maintenance of the project. There are sixteen
people who work with the community house and ten are from a sponsorship group

working directly with children, such as the dance instructor. The house is open during regular working hours and two residents, usually the CDRistas, guard the house at night.

Julia explained that the project co-ordinator works with the different organisations that participate in this community, and with the neighbours. For example, Oxfam Canada supports this project with a Cuban partner organisation called the Centre for Information and Study of InterAmerican Relations (CIERI). She explained that CIERI has supported this project from the beginning, through the process of diagnosis in the municipality and community. Julia continued to explain: "First the CIERI had to identify which members of the community needed the most assistance; it then proposed a report based on these needs". We were informed that the community of *El Comodoro* was only four years old at the time, and that the people living there had been displaced from their own communities. Because of the ongoing housing shortage, they had been living in shelters after their homes had been destroyed by a hurricane. However, relocating people to *El Comodoro* was not the silver bullet to their housing problems. CIERI identified that these people had a number of specific needs once arriving at *El Comodoro*.

Julia elaborated, telling the group that the lack of resources due to the economic crisis has meant that the very basic construction style at *El Comodoro* was the only type possible and that it had to be constructed outside the city due to a lack of available space in Havana. There were also problems with alcoholism, most likely resulting from depression exacerbated by the shortage of transportation to the cultural centres such as Habana Vieja (Old Havana), where many of them had previously lived. Problems with health also emerged. However, the key problem CIERI identified was a lack of feeling of belonging, as many of the residents had come from urban communities such as Centro Habana. In attempting to address these problems the CIERI had advised that the residents construct a community house, which was the house where the presentation was taking place. Once CIERI had identified a problem, members of the community organised the work, placing special emphasis on children in the community – though work with children had begun prior to this in the form of social and cultural activities. Julia explained the process:

> The government supported CIERI to construct this community house. So an action group was created according to local needs and they then allocated work to the people in the community. Unfortunately, the construction of the house took longer than expected because of difficulties getting raw materials. There was an urgent need for community employment. So doing that work brought a determination to people to construct a space for children. Children were the main goal initially because CIERI identified that they were the most affected people in this community. Our mission was to bring cultural construction and integrate the children into the community.

The group was told that, as more community diagnosis is undertaken, other organisations had also been getting involved and supporting the community with numerous proposals presented to *El Comodoro*. For example, a playground and sports area for children was to be created, the location of which would be decided by the children. Julia said that the community also had a goal of improving the primary school so that it would become a more pleasant and comfortable place for them to spend their days. We were told that, like other community workshops and projects, the community house organises workshops in different cultural fields, such as dance, crafts, theatre and so on. After Julia's presentation, the tourists were asked if they had any questions, and almost everyone raised their hand. The discussion below sheds light on the level of interest and the degree of engagement and interaction.

One group member asked how many people live in the community and how many of those are children. Julia responded that "we have 250 families and approximately 250 children up to the age of 16. Many families are older and their children are married and live elsewhere". Another asked whether the school is located in the community. One of the residents explained that there is a primary school in *El Comodoro*, but the high school is a short bus ride away.

One participant asked whether new houses are still being built for people getting married. The response was as expected given everything we had heard about the housing crisis: "One problem in this country is housing, it's a problem not even solved in the capital. So, for them to move from here to the city or to areas from where they originally came the counties would need to construct housing for them. The Historian of the City built some houses for those people wanting to go back to Habana Vieja. But housing is a huge problem in this country. Typically when you get married you live with your parents or move if you can back to your municipality if you can". Questions about the community garden included some inquiring about the size of the garden, which we were told is approximately two blocks and that the main objective of the garden is to create a better dynamic in the community. Another asked whether each person has an allotment, and Julia explained that residents don't get an allotment, but rather people work there to produce for the community and the produce is distributed to the community. She explained that the water for the community *and* its gardens comes from an aqueduct from their old community in Havana.

One group member asked how much the government contributed towards the establishment and maintenance of this community. Julia provided some more context for her response, by explaining "First what you should know is that, in Centro Habana for example, you do not really need a hurricane for buildings to fall down – the hurricane simply accelerated the process of the collapse of buildings. So many people lost their housing in Centro Habana and were living in shelters sharing many things. The government provided the resources to construct small houses to at least provide families with privacy. The rest was undertaken with support from development agencies. For example, a Spanish organisation helped build the construction of this community house, and Oxfam helped finance the the chairs, television, stereo, toys, furniture and so on".

Several questions were asked about employment problems and opportunities for the residents, particularly given the transportation issue. The simple answer was that there are problems. "When the people moved here they lost their jobs especially the women. For example, now 95 per cent of people working in the primary school are from this neighbourhood and this community house also employs only local people", Julia told us.

One former primary school teacher asked Julia about what the community house needed in terms of resources to work with the local children. "Well, we have enough resources to do it. We work with specialists from the House of Culture and are also grateful to the development agencies and human resources of workers. The House of Culture has different instructors dealing with the various arts, such as dance. There's also a proposal to work on ceramics and teach the children about pottery".

"Are there any health facilities here within the community itself?", another curious participant asked. "There are two family doctors with nurses. There is also a ration store here. Our main health problems in this community are high blood pressure, heart disease, alcoholism, and diabetes. There are the typical cafeteria street stalls, but only one that sells alcohol. Alcoholism has led to various social problems in this community, including sex work, and this has led us to realise that there was a major problem with communication in the community, i.e. within families. So we decided that, as part of the project we should act on the communication problems here. We took action to ease communication within the families through activities within the community house. We needed to make people believe they belong to both the community but also to this project. First we aim to encourage neighbours to come to the community house and then they gradually feel that they belong. Among the things we have been doing with children and families are trying to locate communication problems in the family, i.e. parents with children".

After this presentation and the question and answer time, we were invited to wander around the community and meet the local people. Many children had noticed that there was a group of foreigners in the community house and had come to watch. When the group moved outside, many of the tourists chatted to the children. Others used their basic Spanish skills to engage local people in conversation. Once on the bus, I noticed that some of the Australian folk were chatting about the value of such community houses being run by residents and how such a model might work in remote communities of Australia. There seemed to be consensus that while it is great that NGOs are supporting such communities, it is even better to see that it's the local residents who are in control of shaping its activities. Clearly a developing a sense of place has been central to the focus of these activities.

Day Eight – Presentation and Discussion: Education and Literacy in Cuba

Travelling eastward along the south coast to Cienfuegos, we had to drive slowly not because of heavy traffic, for there are very few cars, but because of the annual migration season of small orange crabs (*cangrejos*). There were literally millions of tiny bright orange crabs scurrying from the Caribbean Sea across the highway to their mating grounds and the highway had become a veritable red carpet of crushed crabs. Cienfuegos literally translates to 'one hundred fires'. It is where we had our education presentation and school visit. Cienfuegos was quite distinct from the rest of Cuba, because it was founded by French settlers rather than Spanish conquistadors and is probably the least African of all Cuban regions. As we drove through the streets, we entered into what appeared to be a bustling and thriving small city saturated with decorative neoclassical buildings. Horse drawn carts were as abundant as cars on these streets. We approached the large bay that Cienfuegos is situated on along the southern coast of Cuba. I noticed that there was an industrial sector with clusters of chimney stacks and factories on the outskirts, and the tour leader pointed across the bay to the nuclear power plant. We drove back through the narrow streets to the centre of the city to the provincial house of ICAP (Cuban Institute for Friendship of the People). Here, we were to meet with the Director of ICAP for this particular province, a representative from the Ministry of Education and two local teachers.

We entered into a pale blue and white building with an ornate façade. Pale blue is a very popular colour for houses to be painted, as well as green and pink. There were beautiful decorative tiles on the floors and lovely wooden rocking chairs in the parlour. We were led through to an internal courtyard with white wrought iron chairs set in a circle. Pot plants lined the courtyard and the sun shone across one section. It was a hot morning and we were relieved to see the chairs were in the shade. A lady brought out a tray of tiny cups. The tour participants by now all knew immediately that this meant strong sweet Cuban coffee. The Director of ICAP welcomed us to the province and each of the ladies introduced themselves. It is Milady from the Ministry of Education explained about education in Cuba.

Milady started by confirming what we had already noticed: that the Cuban people are educated and philosophical even in the most informal conversations. This, we were told, stems from an important initiative established at the outset of the Cuban revolution: access to free education. Education, like healthcare, is regarded as a right of all Cuban citizens and is thus provided by the State. It includes all levels of education from pre-primary, primary, secondary, pre-university or technical education, to university. Cuba has attempted to maximise its human potential due to the brain drain at the outset of the revolution, when masses of educated wealthy Cuban people fled to Miami.

"Prior to the revolution" explained Milady, "43 per cent of the population was illiterate and half a million children did not attend school. But, since the revolution, schools have been built, providing universal access to all levels of education and ensuring that children can attend school regardless of where they live or the economic situation of the family". This initially led to enormous decreases in the number of children in the labour force, with the figure ultimately dropping to zero as easy accessibility to schools allowed for increases in the rates of enrolment (Uriarte 2002). In 1960, university and senior high school students, *brigadistas*, formed literacy brigades; this is commonly referred to as the Great Literacy Campaign. Leaving the city and spreading across the country, they aimed to teach every single Cuban to read and write. Two years later, there were 10,000 more classrooms, and by the end of the decade elementary schools had doubled, and the number of teachers had tripled.

Milady explained that a unique feature of the Cuban education system is the emphasis on combining education with work. In fulfilling José Martí's dictum "in the morning the pen, but in the afternoon the plough", Cuban children are expected to be *estudiantes hoy, trabajadores mañana, soldados de la patria siempre* (students today, workers tomorrow, soldiers always). The 'Schools in the Countryside' program started in 1971. Children spend time each summer working in the countryside, where they live in boarding schools attached to plots of arable land. Half of all intermediate school children also attend a rural boarding school, where time is equally divided between study and work usually on citrus plantations, for at least some of their education.

Today, the Cuban government invests 9.8 per cent of its GDP in education, more than twice the average percentage allocated throughout Latin America.[8] The tenth grade is the average educational level reached by most Cubans, and illiteracy rate for Cuban youth between 15 and 24 is zero. The government also focuses on tertiary education and research which is reflected in the continually increasing numbers of university graduates. In addition, there are over two hundred scientific research institutes across the country, which claim to have achieved breakthroughs with vaccines against AIDS and dengue fever, drugs to dissolve blood clots, genetic modifications of plants, fish and animals to promote better yields, resistance to pests and disease. All of this has led to the most highly educated and skilled workforce in Latin America and the Caribbean. The educational attainment of Cubans has resulted in a highly educated workforce. The World Bank observes that Cuban education is outstanding in its achievements of universal school attendance, high adult literacy levels, high female representation at all levels of education, and access to basic educational opportunity across the country including in impoverished areas, both rural and urban (Gasperini 1999: 4-5).

Following the presentation, we were escorted to a nearby primary school, where the children awaited us. One class of fifth graders performed a series of Cuban songs, dances and poetry. Afterwards, while the students were on their lunch break, we were escorted around the classrooms and the principal showed off the school's resources the school has. Some of the group members had saved their donations of pencils and notebooks to give to this school, whereas others had

---

8   Oxfam America.

given their donations to the various community projects we had visited earlier in the trip in Havana.

Day Nine – Presentation and Discussion: Healthcare in Cuba

We then travelled further east to visit the beautiful town of Trinidad. On our way into Trinidad, we drove through fields of sugarcane, passing farmers with oxen, while large trucks lumbered past, carrying people packed into the back. Because of fuel shortages, this is a common form of transport in Cuba and for the same reason farmers have had to revert to working their fields with oxen rather than tractors. We arrive in the colonial city of Trinidad in the morning. We were told by our tour guide that Trinidad has around 50,000 inhabitants and has been declared a world heritage site by UNESCO. Arriving in the bright sunlight of the morning, we noticed the red tile roofs, cobblestone streets, stained glass arches and intricately designed wrought iron grated windows. This is very different from Havanan architecture. Trinidad has maintained its old world charm and elegance through the centuries. Our mini bus dropped the group at the outskirts of the tiny city and we walked through the cobbled streets to the large sunny Plaza Mayor, the central square. Emilio, our tour leader, informed us that the array of statuesque buildings surrounding the Plaza consists of museums and churches. I noticed that this is a popular meeting place for locals, with people sitting on the park benches chatting and watching the goings on. An old man wearing a straw hat slowly walked beside his small donkey as it pulled a cart through the narrow cobbled streets. It felt like we had stepped back in time. After a quick orientation, we wandered at leisure before our meeting on healthcare. There are few tourist hotels in the cobbled traffic-free streets, so most visitors stay in *casas particulares* (private homes). Trinidad offers an unparalleled insight into the close knit community spirit that pervades Cuba because of its size and the ways in which local activity is concentrated.

At the designated time, the group met back at the Plaza, and following our guide, we made our way to the healthcare presentation at the local hospital. As we walked through the hospital, one member of the group was struck by the simplicity of the facilities and wondered aloud if everything she had read about the Cuban healthcare system was fiction. The hospital was dimly lit, mostly by natural light. Inside it was hot and humid; there was no air conditioning and the rooms we passed seemed short of equipment. We were ushered into a room resembling a classroom and the Director of the hospital, a surgeon, and a local doctor took seats at the front facing us. After welcoming the group, the Director informed us that they would tell us about the history of healthcare and the unique healthcare system in Cuba following which we would take a tour of the hospital.

The Director explained that, since the revolution, healthcare ha been viewed as a right of Cuban citizens and is thus delivered free of charge. While the World Health Organisation rates Cuba as one of the highest in Latin America for its level of healthcare services, Cuba did face significant challenges at the outset of the revolution. At the time, healthcare primarily consisted of private polyclinics concentrated in urban areas and a deficient public system. In addition, directly after the revolution commenced, approximately one half of practising physicians fled the country (Uriarte 2002). The Director explained that Cuba was faced with caring for its people with greatly diminished resources and the need to train almost all of its medical personnel. This did however lead to the new Cuban government being able to develop a healthcare system from the ground up. As a result, it has successfully established a system that is internationally recognised for its strengths: universality, reach, accessibility and orientation to preventative healthcare at the primary level. The Director then handed the discussion over to the doctor, who explained that primary healthcare consists of a family doctor and nurse in every neighbourhood who live and work in the community. The doctor is provided with a furnished home and an examining room and receives 400 pesos ($20 US dollars at the time) a month. The family doctor and nurse attend to around 250 families. Family doctors provide primary healthcare in their office and also conduct home visits. They pay particularly close attention to pregnant women, new-borns, children, those with chronic illnesses, the elderly, and those recently released from hospital. The former United Nations Secretary General Kofi Annan stated at the first South Summit in Havana that Cuba's efforts in maintaining high levels of public health care are especially impressive and, according to Annan, what is most extraordinary is that Cuba maintains the same levels of healthcare in the rural areas as in the cities (Madruga 2000).

The Director told us that human well being is often measured internationally by observing infant mortality in a country. The reasons for this are twofold: infant mortality indicates the quality of healthcare available to mothers and new-borns; and it taps into associated variables such as poverty and access to food that affect the health status of mothers and babies.[9] In Cuba, infant mortality rates have been significantly reduced from 35 deaths for every 1,000 live births in 1950 to 7.2 deaths per 1,000 babies born alive in 1999.[10] Other health indicators have also improved remarkably. In 1999, mortality for children under the age of five had reduced significantly to 8 deaths from 54 deaths per 1,000 children in 1960.[11] Child deaths in many developing countries are often due to infectious diseases, but these have largely been eradicated in Cuba through the implementation of immunisation programs and widespread health education. Prior to the 1990s Cuba successfully eradicated a range of diseases such as measles, rubella, typhus fever and diphtheria and had decreased the rate of tetanus and tuberculosis (Uriarte 2002). Child vaccination programs in Cuba rivals those of surrounding developed countries in the Americas, Canada and the United States.

---

9   World Bank (2001) World Development Indicators https://publications.worldbank. org/WDI.

10   UNFPA – UN Population Fund – Country Profile/Cuba.

11   World Factbook 2002 – Cuba www.odci.gov/cia/publications/factbook/geos/cu. html#Ec.

Life expectancy of Cubans has also increased, with elderly people living to an average age of 76.6 years. The types of diseases that most often cause death in Cuba are the same as those in developed countries, predominantly heart disease, cancer, and strokes. The infectious diseases that are the most prevalent causes of death across the Third World have been eradicated and, as a result, the health and life expectancy of the Cuban population has improved dramatically. Many patients, we are told, come from the United States for treatments which are either unavailable or more expensive at home.

Typically, there are a number of medical professionals amongst the tour group members and they are generally very keen to ask about the free health care system. Cuba has a reputation for its AIDS sanatoriums, and as the tour members usually had heard about them either in a negative or positive light, many of the participants want to ask about these sanatoriums and Cuba's low HIV/AIDS infection levels. When one of the group members asked a question on this topic, a murmur of interest rippled through the group. We were told that the sanatorium system was established in 1986, with shelters designed to facilitate treating people. Initially people were made to live at the sanatoriums, because the treatment was intensive and it was a new health problem for Cuba. However, in 1993 the laws were changed and people infected with HIV no longer had to join a sanatorium, and were able to keep their jobs. Interestingly, many people still choose to use the sanatoriums because they provide extensive medical and social support. One role of the sanatoriums is to develop skills to form self-support teams that help sufferers work within a communication network and offer counselling services to one another.

There are also STD clinics that are part of a national program against AIDS operated by the Ministry of Education. These clinics run social communication campaigns to educate the general public but also to train doctors, nurses and students in hospitals. They also conduct research into the cultural response to HIV, and provide counselling and hotlines to deal with questions pertaining to STD infections, means of transmission, matters regarding sexuality and relationship counselling. One of the tourists, Edna, an elderly lady with a PhD in nursing, directed the following query to the surgeon: "I've heard that, although Cuban doctors are amongst the best trained in the world, Cuba lacks equipment and medicines due to the embargo and therefore, because of these shortages, hospitals can often only perform emergency operations". The surgeon responded that he has grown up with the revolution and therefore he believes there is always a solution. For example, if one medicine is lacking then an alternative treatment can be prescribed, such as acupuncture or homeopathy.

Figure 15 A seminar about the National Literacy Campaign

Figure 16 Travelling by convoy into the city of Sancti Spiritus to hear Fidel Castro speak

Figure 17 Farmers now use bullocks for the sugarcane harvest

Debrief

Undoubtedly, a clear message that Cuban people want foreign visitors to leave with is Cuba's capacity to reinvent itself in difficult times. The revolution has proved to be an evolving process and continues to change as it balances re-entering the world economy (through tourism for example) with maintaining a commitment to the values of universality and equity of social services. Faced with a long-standing economic blockade, Cuban solidarity has always existed in the form of support from other socialist countries and international brigades. Now tourism, particularly development-oriented tourism, is yet another way to advance solidarity in a globalising world.

These above annotated and abbreviated snippets demonstrate how tourists are introduced to Cuba and its unique development model. The series of presentations, project visits and pre-tour information is designed to provide a comprehensive insight into contemporary Cuban realities and Cuba's different development paradigm, with its mixture of successes and ongoing challenges. Rather than charting the entire tour itinerary, this chapter highlights some aspects of our visit, with the aim of understanding why we need to consider the presence of tours and the types of interactions they engage in. In doing so, we learn why and how new tourism is taking shape and how the moral underpinnings of these tours lead directly to a place like Cuba.

Throughout the course of the tours people learn how Cuba has endured the economic crisis of the 1990s through the government's commitment to the values of universality and equity that underpin social development. They also learn about some of the measures used to confront the economic crisis that are vastly different from those employed elsewhere in Latin America. Rural poverty, unemployment and inequality have increased throughout most of Latin America, but Cuban government policies and measures have increased production and facilitated economic growth with a more egalitarian result (Saney 2004).

Along the way, however, tour participants learn that despite the continued investment in the social services the quality of services has deteriorated in the last decade. As Cuba struggles to rebuild its economy, it faces critical challenges in the new environment, such as increasing unrest among the Cuban youth. Proof of Cuba's strength, however, lies in its ability to draw on the knowledge and successful experience gained in the process of achieving the high level health status of the population, eradicating illiteracy and maintaining one of the most educated workforces in the northern hemisphere. However, the tourists are provided with a comprehensive pack of readings prior to the tour that presents a selection of views on Cuba, and the dissident voice is absent from the program with the Cuban government monitoring whom the tourists meet. While it is not impossible to meet with dissidents in Cuba, the NGOs typically chose not to include such meetings in the tourism program, so as not to cause individual dissidents potential problems with the authorities and to preserve their own standing. The NGOs do not want to become *personae non gratae*.

Many of the people our groups met with believed that the current social problems facing Cuba, the re-emergence of prostitution, drugs, and crime, are not yet entrenched and are thus amenable to strong interventions. While many countries in the West struggle with their social service systems, Cuba has maintained (in the face of the embargo) a strong social service network, they were convinced is only in need of adjustment to the new global climate. It is these achievements, and the unique ways in which Cuba addresses them, which are of interest to a developing niche of tourists who join small educational development-focused tours operated by international development agencies and human rights organisations.

Mediated encounters on development-oriented tours, such as those described here, provide further momentum towards a new form of tourism. From the perspective of the Cuban organisations, the study tours are an opportunity to promote solidarity and proclaim the participants as 'ambassadors', encouraging them to return home and disseminate what they learn about Cuba to their compatriots. For the tour participants the study tours are an opportunity to support community development work through the money they pay to participate on the tours, learn about another country and its development through 'authentic' encounters with local people and grassroots organisations, exchange ideas with local people, and meet like-minded people. The chapters following in Part 3 present an exposition of the tourists engaging in this form of educational and moral tourism, with the aim of answering some key questions emerging from this research: to what extent are these people agents in a new form of tourism? To what extent is the form of

development-oriented tourism a development initiative or a phenomenon driven
by the values and demands of tourists?

PART 3
Rights-based Tourists in Cuba

Chapter 5

Motivations of New Moral Tourists

> Any anthropological approach to tourism must be based on a thoroughly empirical research strategy which seeks hermeneutic understanding in terms of the knowledge possessed by the participants themselves – their definitions, goals, strategies, decisions, and the perceived consequences of their actions (intended or otherwise). (Wilson 1981: 477)

Nash (1981: 477) tells us that, to address the needs of an anthropology of tourism, a "beginning can be made by questioning tourists". Indeed, within the tourism literature there is "a general lack of empirical engagement with tourists 'in the flesh'" (Desforges 2000: 934). An anthropological approach informs us of people's desires and experiences within tourism and thereby allows us to interrogate how it is that tourism is inevitably an important part of modernity wherever it takes place. Importantly, tourism, in its many forms, provides key insights about modernity because it "is one of the defining activities of the modern world, shaping the ways in which one relates to and understands self and other" (Palmer 2005: 2). Thus, journeys to Cuba to learn about development in the context of socialism might evidence a crisis of values, morals and judgement in Western countries. Perhaps, then, NGO study tours share motivation/desire with broader participatory development that emerges as "consequence of transference onto Third World communities of the perceived inadequacies of our own liberal democratic political systems" (Kapoor 2005: 1208). Essentially, we can learn much about modernity and its value systems through an in-depth understanding of tourist motivations (c.f. Graburn 1989, Palmer 2005). For example, understanding tourist choices offers greater insight into the importance of the leisure ethic or how tropes of sustainability guide morals pertaining to a concern for developmental, environmental, cultural, and social issues at home and abroad. Modernity is most often characterised by comparing modern societies to premodern or postmodern societies; inevitably it needs a chronological framework (Jameson 1990). Although it is possibly too difficult to capture the diverse realities of societies throughout various historical eras, it is accurate to say that the notion of modernity has acquired great fluidity, where it has become plural, uneven, contested and "at large" (Appadurai 1996, 2001).

This section of the book is both empirically and qualitatively based and offers an exposition of the desires and experiences of those people participating in NGO study tours in Cuba. I frame my arguments within the well-established personal transition literature of anthropologists Van Gennep (1960), Turner (1969, 1978), Graburn (1983, 1989) and Nash (1996). These more recent writers suggest that there are historical precedents for modern, purposeful and educational travel, such as pilgrimage, and while the tourism niche that is the focus of this study is hardly a pilgrimage in the sense that it results in spiritual transformation, it does nonetheless offer an approach for analysing NGO study tours.

NGO study tours in Cuba offer tourists an intense learning experience: people return home with an enriched view of development processes. This is where an unmistakeable convergence between development and tourism emerges from broader aspects of modernity's processes of social change and the experience these entail for tourists. To assist in a clearer understanding of the macro-development implications, I sought to undertake a micro-level exploration of what new tourism means through analysing participants' comments collectively. NGOs indicated that they hoped participation in their study tours would permanently change people's commitment to supporting development initiatives on returning home. This concept of transformation has a historical continuity with the proponents of the value of studying transition rituals. As suggested by Turner (1969), and adapted by Graburn (1983; 1989) for the anthropological study of tourism, the analysis of personal transition involves not just the study of the experience and the reactions of tourists but is also concerned with what motivates people to travel and the consequences for them and their home society (Nash 1996: 57).

In NGO study tours, tourism was a two-way exchange where tourists arrived with specific desires and motivations (seeking authenticity, solidarity, learning about development, participating in a form of travel consistent with their personal identity) and left with a more informed understanding of development and an intended commitment to solidarity. At the same time, the tourists felt they had had close personal contact with local people and that the other participants on the tour added substantively to affective elements of their experiences, a process analogous to Turner's normative communitas. Over 60 per cent of participants that I interviewed subsequently considered their experiences not transformative *per se* but rather leading to a deeper understanding of development in Cuba. In turn, they understood their experiences as a re-affirmation of a sense of identity rather than change.

However, I wish to suggest tourism with development and human rights organisations was not just about modes of economic support, but about far less tangible elements that were exchanged and developed, such as the emotional, educational and affective elements that build on identity affirmation. If this was indeed the case, then this style of tourism fitted within rights-based development in which improved well-being for those deemed in need was promoted by broadly-conceived support that was not just about giving money but about creating solidarity through productive synergies of tourism and development. Arguably, the tours contributed to the dignity and well-being of local people because tourists learned about community development initiatives from local perspectives and left Cuba with a strong sense of solidarity that in some instances fostered the creation of international networks. This leads us to the bigger question framing this section of the book, how do we shape our understandings of the ways tourism and global social changes are related?

Modern Touristic Ritual

Purposeful and educational travel such as NGO study tours has antecedents in institutions such as the Crusades, the Grand Tour, and religious pilgrimage. For Graburn (1989), ceremonies, rituals, folklore and, these days, tourism form important symbolic institutions that embellish our day-to-day lives. They exemplify a "human exploratory behaviour". In traditional societies, transition rituals were one such important symbolic institution. This is because, as Giddens (2002: 64) suggests, "tradition is necessarily active and interpretative". Ritual is thus a practical means of necessary social reproduction, as it enmeshes tradition in practice.

Theoretical studies of transition rituals provide important insight into particular niche forms of tourism. Van Gennep (1909; 1960) offers a comprehensive framework for understanding the obligatory social system of transition rituals in traditional, pervasively religious societies, which mark the transition from one status to another, for example from adolescence to adulthood. It is worth momentarily pausing here to problematise any idea that societies can be thought of as traditional as opposed to non-traditional. As Giddens (2002: 57-58) qualifies:

> few people anywhere in the world can any longer be unaware of the fact that their local activities are influenced, and sometimes even determined, by remote events or agencies ... The day to day actions of an individual today are globally consequential. My decision to purchase an item of clothing, for example, or a specific type of foodstuff, has manifold global implications. It not only affects the livelihood of someone living on the other side of the world but may contribute to a process of ecological decay which itself has potential consequences for the whole of humanity.

Thus we can say that most societies have moved into a sort of post-traditional global order precisely because of the accelerated nature of interconnectedness (Harvey 1989; Tsing 2000). No society can be considered static and unchanging, nor can traditions (Shils 1981).

Nonetheless, referring to what Van Gennep defined at the time as traditional societies, the ritual process entailed three important phases: rites of separation from the ordinary, a period of liminality and rites of reincorporation into society. Following on directly from this typology, Turner (1969) showed that ritual processes involve a distancing from the routine of social life. The period of liminality was destructured, non-ordinary and could assume a sacred aura which may involve a state of communitas with those involved in the same process at the same time. The ritual process ended with reincorporation into ordinary structured daily life.

Importantly for the anthropological study of travel, Turner (1977) later saw ritual as not only pertaining to religion but also relevant in largely secular modern societies. He labelled the parallel concept to liminality in religious societies as liminoid for secular rituals. Turner and Turner (1978) tied this term liminoid to the concept of pilgrimage, referring specifically to its voluntary aspects. Through travel, pilgrims were committed to a spiritual centre external to their society; and, in their quest, while voluntary rather than obligatory, they were likely to experience elements of liminas and communitas. The pilgrimage rite involved an existential state leading to a personal transition. This symmetry between obligatory liminal situations and voluntary liminoid situations made it possible to apply this theoretical construct to tourism (Cohen 1988a).

Turner's valuable insights into ritual have been adopted by a number of anthropologists in the study of tourism (Jafari 1987, Graburn 1983; 1989, Nash 1996). Like some traditional rituals, including pilgrimages, tourism involves a separation from normal day-to-day life; a sense of getting away from it all. Graburn (1989) saw tourists entering a non-ordinary life, which has a sacred quality, before coming back to their ordinary existence, which is distinctly profane.

> Thus it has a beginning, a period of separation characterised by 'travel away
> from home'; a middle period of limited duration, to experience a 'change' in
> the non-ordinary place; and an end, a return to the home and the workaday.
> Thus the structure of tourism is basically identical with the structure of all ritual
> behaviour. (Graburn 1983: 11-12)

For our purposes tourism became a "modern ritual" (MacCannell 1976) which directly paralleled the values about health, freedom, nature and self-improvement underpinning pilgrimage and traditional rituals (Graburn 1983: 15). Tourism "offers entry into another kind of moral state in which mental, expressive and cultural needs come to the fore" (Graburn 1983: 11). Development-oriented tours in Cuba are not a form of pilgrimage *per se*; nevertheless, the transition rituals literature, as Nash (1996) points out, offers a particularly cogent framework to inform a study of tourists' motivations and experiences and the implications for development. Such a framework allows us to make assessments of whether tourists undergo a transformation and allows us to assess how tourism is transforming itself into a component of, and with, a larger instrumental role in development. The following discussion stems from observations about what happens on tour and from what I have learned by interviewing and questioning tourists.

Profiling NGO Study Tourists

In order to develop this framework, I shared tours and spoke with nearly one hundred tourists participating in NGO study tours in Cuba. The emergent picture showed that increasingly people who wanted 'meaningful' and educational tourist experiences could purchase them from development and human rights organisations. A central focus of these tours was to educate tourists about the development issues that confronted particular cultures and what local and international efforts were being made to alleviate and improve these conditions. Such tours appealed to tourists who sought to use their leisure time for personal growth and educational purposes. They travelled with NGOs to Cuba to learn about development issues from local people. This generated appeal precisely because incorporating an educational focus within the tourism experience was not typical; tourism generally centres on notions of leisure rather than learning.

By using data from formal interviews, questionnaires, informal conversations and participant observation, I sought to profile NGO study tour participants in order to identify similarities and differences. The typical participant was over the age of 50, female, and tertiary educated. Of those formally interviewed (26 people from across seven tours), 61 per cent were female and 39 per cent were male. The ages of tour participants ranged from 20 to 80, with 18 per cent under 30; 32 per cent under 50; and 68 per cent over 50. Of those respondents who were under 30, all were currently studying at university in a course related to the focus of their study tour. For example, one participant on the Sustainable Agriculture tour was doing a Masters in Landscape Architecture researching urban agriculture. Of those remaining in the under 50 category, all but one were university educated and worked in a field related to the focus of their study tour. Some examples included a permaculture designer and a horticulturist on the Sustainable Agriculture tour, a principal at a special education school on one of the *OCAAT* tours who stated that her primary interest on the tour was education in Cuba, and a nurse interested in the three tiered health care system on another *OCAAT* tour. Of those respondents over the age of 50, 79 per cent were tertiary educated and 63 per cent were involved in work and hobbies related to the theme of the tour. Thus, tour participants are predominantly tertiary educated and/or working in (61 per cent) or retired from (28 per cent) white-collar jobs. This, coupled with the expensive price tag of a study tour with a development or human rights organisation, clearly indicates a middle class demographic, which is consistent with other analyses of new consumers/new tourists (see Poon 1989, 1993). In fact 71 per cent of my informants said they were involved in something related to the focus of their tour. People named a diverse range of activities associated with sustainable agriculture/organic gardening (32 per cent), the environment (21 per cent), education (36 per cent), and women's issues (14 per cent). For example:

> I spend time with an after school youth program called The Boys and Girls Club in San Francisco and one of the programs I run for them is a gardening project. I grow mostly food in the courtyard of our building with the youth in our program. I have always felt that as a society we are very wasteful so sustainable agriculture was a natural point of interest given my perspective. (Ruben – Sustainable Agriculture tour)

> I have a keen interest in gardening. I've worked for Denver Urban Gardens as a farm intern, working on the organisation's organic farm and as a design intern, designing and building community gardens. (Wilhemena – Sustainable Agriculture tour)

Most tour participants had travelled widely prior to their trip to Cuba and just under half had participated in a study tour previously. Furthermore, the majority of respondents claimed to be engaged politically, socially, and environmentally with many involved with new social organizations, which mobilize around issues such as ecology and human rights. Many participants were also involved in philanthropic or environmental activities and gave substantially to charities and travelled widely as a means of learning. For example, one tourist, Gertrude, supported Oxfam in numerous capacities: "I contribute to Oxfam's Aware program, I've belonged to my local Oxfam group for a long time, I participate in their Walk Against Want campaign each year and worked as a volunteer for Oxfam's Taste of a Nation for three years. I've travelled with Oxfam to Vietnam, India, Guatemala and now Cuba".

Of the seven tours that comprise this research, three were *GERT* tours called Sustainable Agriculture, the Women's Delegation, and Cuba at the Crossroads; and four were *OCAAT* tours, which incorporated a comprehensive array of topics including sustainable agriculture, women's issues, healthcare, education and so forth. All tours commenced in Havana and travelled through six to seven provinces over several weeks. Participants attended a busy schedule of meetings and projects everyday, usually with one day free in the middle of the itinerary. In the evenings there were opportunities to immerse oneself in the local environment, by dining in Cuban *paladares* (private restaurant), attending a rooftop *Peña* (dance and music), or just sitting on *el Malecón* (the seaside walkway in Havana) socialising with local Cubans. Many participants commented positively on the mix of learning and leisure on the tours.

Sacred Travellers' Desires

Why is the analysis of tourist motivation relevant to an anthropological study of tourism and development in Cuba? It is important to note here that this research is not an impact study *per se,* but examines the process and the subjects rather than ostensible objects of this process. At its most basic psychological level, motivation is the compelling force driving all behaviour and is thereby a determining variable in tourist behaviour (Crompton 1979: 409-410). In fact, it has been argued that a study of tourism that fails to address tourist motivation is largely an unproductive project (Nash 1981: 470). More specifically, motivations are important elements of in-depth understanding of the study tour experience and hence what implications this might have for development in Cuba. Considering peoples' motives, desires and expectations helps us to examine whether these tourists demonstrated a genuine interest in the development issues of the 'Other',[1] or whether they were more interested in new personal experiences, what Munt (1994b) labelled as

---

1   With reference to development and tourism, see the foundational work: Rosaldo, R. (1989) 'Imperialist Nostalgia', *Representations*, 26: 107-122.

ego-tourists, a criticism often levelled at mass tourists. A genuine interest in the development issues of the 'Other' would indicate potentially positive implications for this tourism-development nexus. On the other hand, if tour participants were more interested in accumulating new personal experiences, as is the case with ego-tourists, the likelihood of positive outcomes for development would be far less because of minimal effort invested by tourists in achieving this outcome. Furthermore, following Kapoor's (2005) line of inquiry, it could be argued that Western tourists are inevitably complicit in the underdevelopment of the Third World, using tourism as a way to satisfy their own desires for leisure and re-creation.

Where pre-modern forms of travel were mostly for religious, educational, medical or trade purposes, Cohen (1972; 1995) argues that motivations of contemporary tourism are significantly different and are continually shifting in relation to changes within tourism. Drawing on the personal transition literature, we can elaborate on Cohen's (1972; 1995) view and propose that, in fact, shifts in modern tourism have not only seen a diversification in types of tourism but a revival in educational travel, which falls under the umbrella of 'new tourism' (Poon 1983). At the same time we can also identify an increasing 'moralisation of tourism' (Butcher 2003). NGO study tours to Cuba are undoubtedly one such example.

Arguably, tourism is one of the principal avenues through which the worldviews of many people are shaped and enacted, both consciously and subconsciously. We might suggest that, on a subconscious level, NGO study tourists are potentially motivated by their 'desire' to be involved in the development of the 'Other'. Such engagement means satisfying a need for deep seated psycho-analytical (Lacanian) resolution through attempts to create a sense of unity, while at the same time denying and thereby ignoring their own 'complicity' in the underdevelopment of the Third World (Kapoor 2005). Kapoor's notions of 'complicity' and 'desire' have not been considered previously in the context of a tourism-development nexus and are useful for arguments that alternative tourism has neocolonial tendencies. Munt (1994b) posits that the 'cultural preservation' agenda is perpetuated for Western bourgeois travellers to make themselves feel more 'worldly', in other words, a process of catharsis. The advocacy of the need to protect cultures has resonance in the colonisation of the past, a kind of racism that celebrates 'primitiveness' (Deleuze and Guattari 1988: 178; 429). But we can further surmise that forms of development that consciously aim to transcend Western-centric values and neocolonial tendencies, such as participatory development, are in fact "prone to an exclusionary, Western-centric and inegalitarian politics" precisely because their actions ensure "the reproduction of inequality and empire" (Kapoor 2005: 1204).

Can development-oriented study tours to Cuba, which grow directly out of the current emphasis on participatory development, thus be seen as a means of overcoming such complicity, or are tour participants subconsciously motivated by a desire to be seen as benevolent whilst simultaneously treating the Third World as object and resource? Or is there in fact something genuinely productive that emerges in a more positive sense from such tourism?

On a more conscious level, tourists are perhaps most overtly motivated by ideas of representation. Indeed, the very nature of tourism is concerned with representation and interpretation (Mowforth and Munt 1998: 7). This relates to place, culture and identity. Representation and interpretation of identity are as much about the tourist as about the destination. Thus the way in which we represent ourselves through our lifestyle choices, including our travel choices – choosing to stay at enclave tourist resorts or travelling to remote regions with niche tour operators – reinforces ideas about our own identity. Hence NGO study tour participants are arguably involved in accumulating cultural capital and thereby might be labelled ego-tourists because they like the prestige associated with travelling with Oxfam or Global Exchange to Cuba. Alternatively, they may have a more moral agenda of assistance and learning which, in turn, they perceive makes positive contributions to development. I wish to assess which of these is the more accurate description, or whether it is indeed both.

Earlier, I showed that certain organisations promote their travel experiences as possessing specific characteristics including a development focus, authenticity, education, responsibility, unique opportunities to meet with like-minded individuals, and the contribution of resources to aid the development activities in the country being visited. Questions concerning the motivations of tourists participating in NGO study tours, the nature of study tours and their implications for development become significant through the degree to which promotion of such characteristics is effective. The tourist's conscious "inclinations at the outset will tend to condition all of their experiences and reactions. This suggests that any use of the paradigm of tourism as a personal transition must begin with the tourist's initial inclinations at home" (Nash 1996: 49). In order to gain insights into this personal realm of individual inclination, I asked tourists participating in study tours a series of questions concerning their reasons for taking a study tour with a development or human rights organisation. A series of themes emerged around motivation that was not mutually exclusive.

The interviews, questionnaires, and participant observation showed that motivations for joining NGO study tours in Cuba covered fairly conventional terrain: to have authentic encounters; to have opportunities to learn, particularly about development; to support Cuban solidarity; and to participate in a form of travel that is consistent with personal identity. Many of the tourists reported that they wanted to experience the 'real lives' of people from other cultures, thus indicating that they associate NGOs with authentic experiences of Cuba. Participants cited a desire to learn about development in Cuba because they perceive that knowledge of such issues improves understanding and appreciation of other cultures. They also said they wanted to travel with NGOs because they knew a proportion of the money for the cost of the tour went to the community projects they visited. Participants indicated that partaking in this style of tourism is consistent with other aspects of their 'alternative' lifestyles and reflected demonstrable and desirable aspects of their identity. Meeting like-minded people was another reason that was cited as a motivation for participating. Responses indicate that developing friendships, a sense of belonging, and experiencing social interaction with like-minded people on NGO study tours gives participants a sense of place and meaning. Many referred to an intellectual objective of participating on these tours; they wanted to learn about Cuban development issues. This is significant because incorporating an educational focus within the tourism experience is not typical of tourism in general. Some participants expressed a strong desire to support Cuba in the face of a US-imposed economic trade embargo.

The Search for Authentic Encounters

Participants in this study associated NGO study tours with the chance to experience 'real' or 'authentic' Cuba. The construct of authenticity operated on a number of different levels but also related to and overlapped with the other tourist motivational themes here of education and development. Due to the complex and vexed nature of authenticity, there has been extensive debate in the anthropological literature, with early anthropologists such as A.P. Elkin and W.E.H. Stanner focusing on salvage anthropology and interrogating notions of 'real' culture.[2] Indeed, authenticity has emerged as something of a trope within anthropology. Not surprisingly, arguments about authenticity have extended to the tourism literature, regarding the extent to which a search for authenticity motivates modern tourists. While authenticity is not applicable to all forms of tourism, because many tourist motivations or experiences cannot be explained in terms of seeking authentic touristic experiences, it is particularly significant to NGO study tours in Cuba.

It is useful to briefly map three central tenets of this dimension in the study of tourism, namely those of Boorstin, MacCannell and Turner, in order to apprehend how authenticity is perceived and experienced by the participants in this study. Boorstin's (1964) study of human experience in contemporary America drew on the modern mass tourist experience to argue that people do not experience reality, but instead indulge in 'pseudo events'. His principal concern was with the contrived and illusory nature of the contemporary human condition. Jean Baudrillard's (1988) notions of simulacra implicitly support Boorstin's earlier ideas about how we 'experience reality'. For Baudrillard and Eco (1986: 6), Americans (and, by extension, Westerners) construct imitations of themselves where "the past must be preserved and celebrated in full-scale authentic copy; a philosophy of immortality as duplication". Thus, the reproduction is more real than the original and the

---

2   For a more comprehensive review of authenticity literature some examples include: authenticity and Aboriginality – Andrew Lattas (1993); authenticity and Aboriginal art – Eric Michaels (1994); authenticity and the Pacific – Jocelyn Linnekin (1983), Margaret Jolly (1992), Robert Norton (1993); authenticity and tradition – Nicholas Thomas (1992), Eric Hobsbawm and Terence Ranger (1983).

simulacrum becomes the true. "The true (like the real) begins to be reproduced in the image of the pseudo" (Morris 1988: 5).

In the context of tourism, Boorstin (1964) argued that tourists are detached from the local culture essentially because they travel *en masse* in guided groups, thereby disregarding the real world around them. He argues that tourists willingly indulge in the inauthenticity of contrived attractions. While Boorstin's analysis sheds light on the origins of postmodern notions of hyperreality, he cannot offer a penetrating analysis of modern tourism, given that he wrote at the dawn of popular jet travel, the technical enabler of mass tourism overseas.[3] Like Baudrillard and Eco, Boorstin made generalisations about Americans, failing to recognise the diverse and manifold motivations and experiences of tourists and constructed meanings within touristic experiences. Some authors (c.f. Bruner 2005) have since rejected concepts such as simulacra and pseudo, partly because what is presented in a tourism context is constructed specifically for tourist consumption.[4] For example, with regard to tourism performance, Bruner (2005: 5) argued that they emerge "from within the local cultural matrix, but all performances are 'new' in that the context, the audience, and the times are continually changing". In fact, Bruner argued that all cultures are continually reinvented and we should therefore be aiming to transcend dichotomies of original/copy and authentic/inauthentic because, as established by Derrida (1974), either/or binaries are well established in Western metaphysics, where one term is often privileged over the other (2005: 146). It usually implies that the original is better than the copy.

MacCannell's (1973, 1976) ethnography of the social life of modernity commented on the crisis of contemporary life in terms of authenticity as manifested through tourism. Where Boorstin disparaged modern mass tourism, MacCannell focused on the quasi-pilgrimage and motivations to experience the sacred. He drew on the long established critique of Western civilisation – alienated man in search of self – in his position that modern tourists, dissatisfied with their lives, sought some sense of authenticity elsewhere. The inauthenticity of modernity was counterposed with the sacred in traditional society (1973: 589-590). For MacCannell, "sightseeing is a form of ritual respect for society and tourism absorbs some of the social functions of religion in the modern world" (1973: 589). Thus, for Boorstin, the modern tourist represented the inauthenticity of modernity; whereas, for MacCannell, the tourist embodied the middle class search for authenticity, becoming the pilgrim of the modern secular world (MacCannell 1973: 589-593). Drawing on the work of Durkheim, Levi-Strauss, and Goffman, he argued that, through tourism, people sought to move beyond the inauthenticity of their lives, travelling beyond the spatial and temporal boundaries of their every day lives to experience the reality of other cultures.

---

3   The influence of that 1959 turning point tends to be under-estimated in tourism studies.

4   Alternatively, Feifer (1985) and Urry (1990) claim some tourists, referred to as post-tourists, reject the realist genre and notions of authenticity.

Drawing parallels between Turner's study of religion and pilgrimage (1973) and MacCannell's study of tourism, Cohen (1988a) utilised the concepts of 'the Centre' and 'the Other' and devised his own typology of modes of tourist experience. The principal thrust of Cohen's adaptation was that modernity alienated some people, who went in search of authenticity elsewhere. Thus, people not alienated from the Centre of their own society pursued travel experiences unconcerned with the 'authentic' and thereby reminiscent of Boorstin's representation of the tourist; while those tourists who felt alienated from the Centre of their own society sought alternatives, embracing the Other, and even turning the lives of the Other into their own elective Centre. From this perspective, Cohen argued, these tourists, referred to as existential tourists, looked for existential experiences embedded in authenticity. Cohen's existential tourist type transcended MacCannell's conception of tourists who sought authentic experiences in the real lives of the Other but did not adopt it as their elective Centre. In Cohen's typology, MacCannell's tourist was referred to as experiential (Cohen 1988a: 35-36).

The people I travelled and talked with were not, as MacCannell said of all tourists, living shallow lives alienated from their society and thus seeking authenticity elsewhere in the 'real' lives of others. My profile of the participants in this study indicates they were affluent, middle class people with a strong sense of who they were and their place in the world. They were clearly travelling at a point in their lives when their time and finances allowed them to do so. For them, tourism was consumption and a study tour with NGOs in Cuba was an expensive and ideological status marker (Bourdieu 1984). I will come back to this notion of identity later, as it forms another key motivation. It is important to highlight here, in relation to arguments of authenticity, that some tourists actively chose styles of tourism that facilitated people-to-people contact, perceived to be more authentic and thus more meaningful than other forms of tourism and therefore considered morally superior to mass tourism.

"MacCannell's work is based on a deep structuralism which fundamentally looked beneath the surface to an assumed real underlying structure" (Bruner 2005: 5). This position drew on Goffman's (1959) conception of front and back regions in order to elucidate the problem of authenticity and the touristic quest for it. MacCannell argued that the modern tourist searched for authenticity but was often presented with "staged authenticity". By this, MacCannell referred to a constructed front region that disguised the real back region, to which tourists did not have access. In his view, the authentic culture was always located at the back away from the tourist gaze (Urry 1990). The frontstage was a constructed response from those subjected to the penetration of the tourist gaze. For many tourists, 'getting off the beaten tourist track' was a way of expressing their desire for entry into the back region and their perception of the front regions' inauthenticity.

Some respondents expressed their desire "to travel more widely in Cuba, hopefully avoiding tourist traps like Playa Ancon and Varadero" (Francesca – Oxfam tour) and "to satisfy my interest in destinations off the 'beaten' tourist trail" (Stan – Oxfam tour), indicating a search for authenticity was a central motivation.

Crick however challenges this notion of staged authenticity (the front region) stating that "all cultures are 'staged' and are in a certain sense inauthentic. Cultures are invented, remade and the elements reorganised" (1988: 65-66). Thus for Crick, staged authenticity is essentially part of all cultures as they continually re-invent themselves. Arguably, the experience of authenticity is pluralistic because it results from individual subjectivities and interpretations. When I was travelling with tour groups through Cuba, I listened carefully to participant conversations and was often involved in them. Their discussions shed light on the nuances underpinning perceptions of authenticity which were reflexive and often critically interpretative (Bruner 2005), as the following discussion will illuminate. It is likely that this was partly a result of the pre-tour literature provided by the NGOs and of the participants' high education levels and prior knowledge which disposed them to pick an NGO tour in the first place.

The high percentage of participants motivated by a desire for an 'authentic' experience indicated they believed NGOs well positioned to provide this experience by allowing participants to go beyond the 'front stage' (Goffman 1959; MacCannell 1973; 1976) and meet people who they might not meet otherwise. "I'm travelling with Oxfam because I felt it would possibly offer me contact with and insight about the country, its people and their culture and politics" (Amelie – Oxfam tour). Innes, also from an Oxfam tour stated: "I previously enjoyed going to Vietnam with Oxfam and travelling at a fairly low level, which enabled me to have a greater contact with the locals and to get a real feel for the country and that's one of the big pluses of *OCAA* tours".

Referring to the itinerary of meetings and project visits, Grace, who joined a Women's Delegation, said: "I was really interested in hands on contact with people. I've always been interested in women's issues ... health and ... care of children". Another participant in a *GERT* tour, whose parents were part of the Cuban diaspora in Miami, said: "we wanted to see how the people really lived under the Castro system" (Delila – Cuba at the Crossroads tour). A participant from the Sustainable Agriculture tour with an itinerary of meeting local *campesinos* and visiting sustainable farming co-operatives said: "I was interested in taking this type of tour because of the hands-on experience" (Josephine – Sustainable Agriculture tour).

The desire to have authentic touristic experiences indicated that people felt this type of tour was morally superior because they perceived that they experienced the 'real' Cuba to which other tourists do not have access. But they also believed it was a 'better' experience because they used the tours as a means to satisfy their own needs. These perceptions related to Kapoor's (2005) notion of complicity and raised the question, what precisely was the 'goodness' of this style of tourism in a moral sense? As we will see in the next chapter, notions of goodness and morality emerged out of their experiences. "The tour looked as if it would give us the opportunity to see aspects of Cuba we wouldn't get to see on a normal trip and to make some real connections with local people" (Tom – Oxfam tour). Examples of participants' perceptions of authenticity implied that people joined NGO study tours because they saw the role of NGOs as being able to offer closer contact to a culture than might be available through mainstream tour operators or by travelling independently.

There was a strong sense from those in NGO tour groups that Cuba was unique because of the situation the trade embargo had created, where Americans, and American products, were are not pervasive in Cuba. It was seen as a country that straddled the development fence in that Cuba had been largely successful in social development but remained largely untouched by America and large tourist developments. 'Authentic', for the participants in my research, meant seeing Cuba while Castro was still alive, before hordes of American tourists (in particular) entered, and while tourism was still relatively contained:

> There seemed a sense of urgency to get to Cuba before the Americans had a big impact on the tourist scene and before Castro died; I wanted to experience the real thing and thought that would take a long time to do, if even possible, if I went by myself. (Ingaberg – Oxfam tour)

> It's an interesting phenomenon socially and politically. I wanted to see for myself. As I said before, I wanted to see Havana in particular, before it all changes and possibly becomes over-developed. (Amelie – Oxfam tour)

> We wanted to get there before the Yanks were allowed back. We've always wanted to go to Cuba. (Tom – Oxfam tour)

The number of participants claiming that they wanted to see Cuba for themselves indicated a level of scepticism about the veracity and integrity of US government and media discourses about Cuba, which were largely weighted in opposition to the Cuban government and its policies:

> So much of the information we get about Cuba in the US comes from one extreme or another, I wanted to try to see for myself what's really going on, especially in areas like human rights and justice. (Henrietta – Cuba at the Crossroads tour)

> The US and Cuba have had a love-hate relationship for many years and I was interested in seeing for myself what Cuba was like … a desire to see for myself what the situation in Cuba is. (Valmia – Cuba at the Crossroads tour)

> We wanted to see how its version of socialism was working out and to find out more about the little island which has been such a thorn in the US side for so many years. (Tom – Oxfam tour)

> My knowledge came from consuming newspaper and other media reports over a long time dating from the revolution and subsequent actions including Soviet involvement and withdrawal. I had assessed that there was something special

about a Cuba essentially isolated by US sponsored blockade which survived even after the Eastern Bloc assistance evaporated. I wanted to check out my impressions and assessment. (Stan – Oxfam tour)

Comments such as these signalled that people trust NGOs to act as vehicles for learning and to provide opportunities to acquire firsthand, 'authentic' information about Cuba. This leads to questions about what participants saw as the main function and role of NGO study tours. People not only saw NGOs as organisations working in the field of development and therefore positioned to provide authentic tourist opportunities. Providing authentic experiences was also important for NGOs' ability to use this perceived sense of veracity as a platform from which to build activities, or make projections for subsequent communication and development goals.

Development

Firsthand experiences of development also motivated people to join NGO study tours to Cuba.[5] This related to the theme of authenticity, in that people felt that, by understanding Cuba's development issues, they had a far more authentic experience and this in turn contributed to their moralisation of this experience. The NGOs in my research promoted their travel experiences as possessing two specific characteristics that pertained to development: participants had opportunities to contribute to the NGOs' development activities and to learn about development because tours were development-focused. There was a common perception amongst the tour members that, by travelling with NGOs, they were contributing to the development efforts of NGO aid programs. Both Amelie's and Tom's statements reflected this sentiment: "I was very happy to support *OCAA* by going on their tour" (Amelie – Oxfam tour); "I liked the fact that a portion of the money we paid for the tour went to the organisations we visited" (Tom – Oxfam tour).

Through tourism programs, NGOs taught tourists about the development challenges faced by the people of the country visited and what efforts were being made to address them. NGOs hoped that tourists would return to their own communities and disseminate the new knowledge acquired on tour, as this should have the potential to have a significant impact for development. Increased understanding would result in changed attitudes and behaviours leading, in turn, to a more just and equitable relationship between developed and developing nations (D'Amore 1988) and, in turn, to successful development. I will discuss this claim and how it is taking place in Cuba with these NGO tours, in the Solidarity theme later in this chapter.

---

5   Over 80 per cent of participants indicated that participating in a NGO study tour was motivated by a desire to learn about development initiatives in Cuba.

The various development foci of the NGO study tours were of particular significance to the motivations of the participants. A specific desire to see first-hand some of Cuba's so-called achievements in social development, as alternatives to the American or Western system, was a clear motivation for many participants, as Maria, a participant from Brooklyn on the Women's Delegation pointed out:

> I chose this particular tour with Global Exchange mainly because it was one that emphasised more health which is always interesting to me and especially because Cuba is supposedly doing so much in the area of healthcare and we who are privileged in the US pay for health insurance and still have very poor care … ah I wanted to really look at how healthcare was really nurturing people here in the community.

Likewise, many participants were keen to experience Cuba's model of sustainable and organic agriculture. For example, Oscar claimed:

> my initial interest was about their success with organic and small scale/urban agriculture … I wanted to learn from a people who succeeded using organic agriculture. The corporate/chemical agriculture group in the US tries to claim that organic can't work to feed people and other bogus slander.

And Sebastian joined the Sustainable Agriculture tour:

> because of the agriculture theme and the organisation it was sponsored by [Global Exchange] … I believe that sustainable agriculture is important, particularly needed in the US where land is over farmed in an industrial style. I also found the idea interesting that this was coming not from a highly developed country but [from one] which was struggling.

Other participants simply wanted to visit development projects as a way of meeting local Cuban people, as expressed by Francesca and Gertrude respectively:

> I wanted to see what sort of projects CAA was supporting and to meet the Cubans involved in these projects … to demonstrate my commitment to CAA … to demonstrate my solidarity with Cubans.

> I had travelled on similar tours to Vietnam, Guatemala and India and found that they provided opportunities to visit local government and NGO projects, schools, hospitals, women's groups etc … understanding that it would enable me to meet local people in a way I couldn't as an independent traveller.

We see clearly that some tourists wanted to learn about development issues as a way of understanding culture and that potentially those people would return home and disseminate their new-found knowledge. This, in turn may lead to changes in behaviour, such as a greater commitment to supporting the development efforts of the NGO, discussed in the next chapter.

Participation in NGO study tours were perceived to be more meaningful to tourists than other forms of tourism, fitting neatly with their lifestyles and broader educational and environmental agendas. For example, most people I spoke with were philanthropically, politically, socially and/or environmentally engaged. The above accounts from the tourists give us the sense that they saw *OCAAT* and *GERT* tours as means for learning about key development issues in Cuba. The tours therefore appealed to their desire to learn while on holiday. This ushers in the next theme, which intersects with both authenticity and development.

Educational

The desire for some educational elements in holidays traverses the themes of authenticity, development and identity because people I questioned felt that learning about key development issues provided a more authentic touristic experience and reflected participants' interests. The NGOs aimed to educate people travelling to Third World countries about the development issues specific to that of those countries, about fair trade and about other 'responsible' travel issues that ensure minimal negative impact on the local environment and people. It thus allows the maintenance of the self-perceived moral agenda.

By choosing NGO study tours, people felt they were demonstrating a commitment to good causes and learning about other cultures in ways not accessible to mainstream tourists. Such sentiments were consistent with other analyses finding that travel can be motivated by a desire to learn (Crompton 1979, Krippendorf 1987, Roggenbuck et al. 1990, Weiler 1991). Participants indicated that educational travel provided opportunities for personal growth and development:

> I figured it would offer me wonderful insight and views on the world and the
> US itself. Developing countries offer me a view of the real world, a world that
> the majority of humans live in, not the exotic Disneyland that the US is. (Oliver
> – Oxfam tour)

The nature of the tourism experience in NGO study tours meant, for some people, developing a greater overall awareness of the world, and perhaps even a sense of place, by participating in a holiday with a purpose (other than leisure and relaxation). In fact, some people joined NGO study tours with specific goals in mind, Stella, a participant on the Women's Delegation, said: "I have been doing research on Cuban women, so that was one reason I came on this tour" and Wilhemena, a participant on the Sustainable Agriculture tour, said she joined:

for its urban agriculture programs which I learned about through my own research of urban agriculture ... I had received grants from the University of Michigan and a landscape architecture honour society to research urban agriculture in Cuba ... when I learned about Global Exchange and found that they had a tour focusing on my research topic I decided to sign up.

In fact, over three-quarters of participants indicated that they were partly motivated by an objective of learning about something specific to Cuba, for example, sustainable agriculture, healthcare, and women's issues.

Many participants wanted to learn about Cuba's development achievements, to take this knowledge home and put it into practice in their own communities. This is important precisely because of its centrality to solidarity and the goals of the NGOs. Henrietta was particularly interested in learning about the socialist system and healthcare:

I'm interested in seeing what socialism can do and what we can learn from its successes and failures. Cuba seemed a good place to do this; maybe the only place ... I wanted to learn more about the healthcare system there, to see if I could bring back first hand information to inform my speaking and writing here.

Others wanted to learn about Cuba's organic revolution in order to return home and implement some of the methods local Cuban people are using in the city. The creative ingenuity of Cuban people saw the local response to food shortages during the Special Period experiment with urban agriculture. It has evolved into a highly developed form of urban farming with half of the fresh produce consumed by over two million Havana residents grown in abandoned lots wedged into the crowded topography of the city. Ruben from the Sustainable Agriculture tour said that:

I wanted to get a close look at the infrastructure set up in Cuba to increase its organically grown food outputs ... I was hoping to increase my knowledge of organic farming so I could bring back useful skills/ideas for the project I work with in San Francisco.

Oscar from the same tour said "I wanted to bring back any agricultural and cultural awareness I could share with people at home about feeding their families from their own gardens". Statements such as these showed that the NGOs' objective, to provide tourists with information about development in Cuba, which they would disseminate once they had returned home, was being achieved to some extent.

Respondents were interested in politics, too. A significant number wanted to experience and learn about socialism: "I was interested in Cuba as one of the last communist strongholds in the world" (Innes – Oxfam tour); "I am interested in socialism and Castro" (Ingaberg – Oxfam tour); "we are interested in the social experiment of the past fifty years" (Bain – Sustainable Agriculture tour). Many participants were concerned with learning about *el bloqueo* too. For example, "Cuba interested me in particular because of the US embargo and its effect" (James – Oxfam tour) and:

> I felt the embargo however horrible to live through created an amazing situation for Cuba. All of a sudden Cuba was forced into increasing its sustainability. I thought Cuba was an excellent case study as far as a country's ability to live in a more sustainable way – environmentally speaking as well as politically and economically. (Ruben – Sustainable Agriculture tour)

Statements like this highlight the sanitised version of social, economic and political aspects of life in Cuba presented to tourists. NGO tourists were generally not looking for problems within Cuban society; rather, they travelled to Cuba out of a sense of sympathy (c.f. Hollander 1981). Kapoor (2005) gives us a framework by which to critically question the sympathetic interest that tourists had in Cuba, and their ideas that they were participating in a better tourism, or 'morally superior' tourism, as Butcher (2003) would have it. Tourists on NGO study tours sought exposure to alternative politics as a way (consciously or subconsciously) of grappling with Western complicity; but, by not challenging their government's foreign policies in regard to Cuba, they reinforced their own complicity, thereby ensuring "the reproduction of inequality and empire" (2005: 1204). However, we cannot ignore that the emergence of new social movements is a cogent response to the unaccountability that underpins neoliberal economies. A wave of political dissatisfaction drives some tourists (and indeed development practitioners) to places like Cuba in search of alternative models to resolve their discontent at home.

Participants indicated that they prepared extensively for their trip by reading pre-tour notes provided by the NGO, widely reading literature about Cuba and watching Cuban movies. These statements might indicate the tours did not fundamentally change attitudes because participants demonstrated a keen awareness about Cuban issues prior to their tour. Their statements signalled that people already had, or were sympathetic towards, the attitudes desired by the NGOs and that they wanted to learn more about development issues in Cuba. Perhaps then the tours simply reinforced the tourists' convictions about their beliefs, identities, and worth as modern citizens of the world. This raises important questions about experiential learning and whether such tours have the capacity to provide opportunities for personal transformation; and, if so, what implications they have for development in terms of tourists as agents of change. I will turn to these questions in the next chapter.

Solidarity

Among the positive social and cultural benefits attributed to tourism is the promotion of goodwill, understanding and peace between people and nations (Var

Ap and van Doren 1994). This view argues that tourism provides the opportunity for people to understand other customs and to exchange information and ideas. The role of tourism as a vehicle for international understanding has been accepted by NGOs such as Oxfam Community Aid Abroad and Global Exchange. Echoing President Kennedy's speech during the establishment of the Peace Corps, Var et al. (1994) contend that this motivation provides a setting conducive to the development of harmonious relations and world peace. The suggestion that tourism promotes understanding assumes that much of the harm that has been perpetrated by Westerners on the people of developing nations has been and continues to be based on ignorance (c.f. Chambers 2005 on 'immersion' as the required corrective to development ills).

> ... tourism has been recognised to be an instrument of social and cultural understanding by the opportunity offered to bring different people into contact and to provide facilities for acquisition and exchange of information about the way of life, cultures and language and other social and economic endowments of the people as well as a chance for making friendships and achieving goodwill. (Kaul cited in Var et al. 1994: 29)

While the NGOs did not explicitly identify solidarity as a study tour outcome, it was implicit in their objectives: to provide tourists with opportunities to learn about development issues and return home more committed to supporting Cuba's development efforts. Solidarity through tourism could be considered an important tool for development agencies, social movements and NGOs in terms of new and explicit ways of promulgating issues of rights, social justice and good governance. In this way, solidarity had a formative and discursive relationship with rights-based development. Solidarity became important in the tourism context, because it was implicitly expressed as an objective of NGOs and explicitly expressed by tourists as a key motivation for participating in NGO study tours.

Effectively, we might think of solidarity as a means for tourists to act as agents of change in the development process. Indeed, it is a novel way Cuba has developed for partly transcending the economic and social constraints of the blockade. For decades, Cuba has received Friendship Brigades as a form of solidarity from all over the world. Teams of volunteers visit Cuba, staying with local families while working in the countryside, usually on orchards.[6] NGOs similarly recognised the value in introducing tourists to their development work through local organisations and the effect this might have for Cuban political solidarity. Development-oriented tours acted as a new form of global coalition and interconnectedness, building on previous alliances, such as with the Soviet Union and other politically sympathetic nations with which Cuba engaged in cultural exchanges. According to Chambers (2005), the practice of immersion gives tourists:

---

6 Friendship Brigades and Volunteers were also characteristic of early Socialist China and Soviet Russia tourisms.

the opportunity to spend a few days hosted by a poor family or community, sharing some of their life, helping them in their daily tasks, learning their life histories, and seeing things from their peripheral perspective ... immersion has shown a potential to make a radical difference to those who can make a radical difference ... A radical reconfiguration of development studies [and tourism] would then include more individual reflexivity, especially self-critical epistemological awareness, and deliberate efforts through practices such as immersions, to gain the experiential learning of reversals.

Some US participants expressed a strong resentment that their government dictates whether they can legally travel to a particular country:

It was a particularly interesting time to go to Cuba because the former President, Jimmy Carter was also planning a trip to Cuba when I would be there. This was very big in politics and not supported by the [then] current President. It made the idea of being in Cuba seem like I was courting danger ... I knew that it was 'wrong' for me to go there ... I was curious, why couldn't I go to a country only ninety miles off the coast of mine ... my motivation was a little rebellious. (Ruby – Oxfam tour)

Additionally, some Australian and British participants expressed negative opinions regarding the US trade embargo and were in part motivated to travel to Cuba as a means of experiencing Cuba for themselves, such as Stan who wanted "to experience a snatch of life under a regime labelled 'repressive communist' by sections of Western media" and James who said: "I had a sense of injustice re the US embargo and wanted to see how the Cubans were coping". Some people such as Kingston and Valmia saw their tour as a means of forging links between the US and Cuba in the face of an unjust embargo:

I was interested in going to Cuba because it has a stirring history and is the object of irrational and unremitting hostility from the US government to a degree unique among nations.

I think the Global Exchange tours are primarily committed to fostering friendship and understanding between the two countries rather than running a commercial business as a mainstream tour operator would.

The theme of solidarity resonated with the goals of the NGOs. The responses quoted above suggested that the participants in NGO study tours were politically aware and engaged and used the NGO study tours as a means not only to learn more, but to support Cuban development in the face of an economic trade embargo. This style of tourism was then, arguably, an advance in creating new spaces in which to oppose neoliberalism, similar to the ways in which new social movements mobilise. An expression of resistance was taking hold globally and tourism was one arena where this was expressed and where resistance was enacted. In essence, these tourists were motivated to be part of a large scale global movement acting as a complex coalition resisting neoliberal globalisation – a kind of Trojan horse. I shall return to this in the next chapter. This leads me to the final motivation theme drawn from the research.

Identity

Throughout the 1970s the anthropology of tourism focused on identity. Valene Smith's (1978) edited collection critiqued the impacts of tourism on the cultural practices which shape collective identities. More recently, Munt (1994a) referred to tourism consumption as a means for defining social identity. Likewise, for Desforges (2000: 929), concepts of identity in tourism studies "point towards the importance of the sort of person that tourists want to become". An examination of the nature of tourists' motivations reveals much about the core value systems of modernity. Graburn (1980: 64) argues that:

> If we are to study the nature of solidarity and identity in modern society, we cannot neglect tourism, which is one of the major forces shaping modern societies and bringing (and changing) meaning in the lives of the people of today's world.

As working and living conditions changed, so did travel preferences. Graburn (1983: 24) contended that:

> Changes in tourist styles are not random, but are connected to class competition, prestige hierarchies, and the succession of changing lifestyles, as well as external factors such as the cost and mode of transportation, access to regions and countries, and the state of the economy.

These changing circumstances favoured a tourism emphasising the socio-environmental context and focusing on the 'humanisation' of tourism with descriptions such as 'people-to-people exchanges' (Krippendorf 1987).

For some tourists, holidays were an opportunity to invert their class-based roles. For example, Gottlieb's (1982) and Lett's (1983) studies demonstrated the relationship between tourism and identity. Gottlieb's study of American tourists found that the inversion of identity often played out through tourism choices where people enacted "peasants for a day" or "Queen (King) for a day" (1982: 173). Likewise, Lett (1983) found yachters on charter holidays in the Caribbean invert their everyday behaviour to become more playful and free. Tourism has frequently been cited as providing opportunities for people to escape their day-to-day lifestyle (Gray 1970, Dann 1977, Crompton 1979, Dan 1981, Krippendorf 1987, Pearce 1995, Swarbrooke 1999, Dann 2000). This is reinforced by literature on transition rituals, where tourism is equated with a liminoid period involving a separation from day-to-day life. However, the majority of participants in my study revealed that they saw tourism as an important extension of the lifestyle choices they made daily, and that the type of tourism they engaged in and supported (i.e. responsible, alternative, individual, educational) reflected their lifestyle and acted as a re-affirmation of identity. The style of tourism they participated in and their day-to-day activities combined "in the ongoing construction of the 'social and moral order' in accounts of [their] activities. Descriptions of 'located activities' make inferentially available notions of 'good' and 'bad' visitors and are thus infused with moral nuances" (McCabe and Stokoe 2004: 606).

As we can see from the following, consumption was partly what constituted identity (Wilk 1996) – in this case, the consumption and rejection of styles of tourism. Some anthropologists claim the consumption of commodities is so pertinent that it should form the core area of anthropology (Miller 1995). New consumers used tourism as a means of affirming their individuality. Crompton (1993) drew parallels between the rise of the new middle classes and the growth in consumer capitalism and its emphasis on lifestyle, including the places people chose to live, the activities they engaged in and the holiday choices they made. For example, Stella, a participant in the Women's Delegation, discussed the significance of women's issues to her identity, highlighting why she chose a study tour to Cuba:

> I've been involved in women's issues since I was twenty and so this was something that is important, you know. I sing the women's cause and I've been involved in a variety of groups over the years so that's important to me ... this just seemed to be, ah, more along the lines of where my interests lie.

Likewise, Francesca indicated:

> I've always been an 'armchair' socialist (not a party member) and supporter of Allende, Ortega, Castro and other Latin American left wing leaders who try to help the poor ... nobody would be surprised to know I'm visiting Cuba (again).

NGO study tours could be seen as catering to a middle class identity whose bearers sought to differentiate themselves from the mainstream in part through their tourism choices. The literature argues that new consumers have acquired changed values and lifestyles. Like Poon (1989), I define 'new' tourists as distinct from tourists of the past who preferred mass organised tours; participants also contrasted themselves to mass tourists: "mainstream tours are superficial, infantile, and thus deadly boring" (Kingston – Cuba at the Crossroads tour). 'New' tourists, Poon argued, were more flexible, individual, independent, far more experienced and educated and concerned with environmental and global issues. Many so-called 'new' tourists saw their vacation as an extension of their day-to-day life, with holiday style and destination choice often a reflection of their identity: "I wanted to connect with people in another country who do what I love to do, grow organic food in community. I don't do mainstream tours; never have, never will" (Oscar – Sustainable Agriculture tour). Statements from these tourists were saturated with implicit moral judgements.

Munt (1994b) argued that 'new' tourists were engaged in a hegemonic struggle for cultural superiority with each other in their search for an alternative to mainstream tourism. This involved being able to claim that they travelled responsibly with NGOs. Following MacCannell, this struggle to distinguish themselves from other tourists was in essence an attempt to accumulate cultural or symbolic capital (Munt 1994b; Urry 1995; Clifford 1997; Holmes 1998; Desforges 1998, 2000). This was an important aspect of the 'new' tourist. Typically, the tourist accounts were laden with moral evaluations about their tourism endeavours and "the ways in which they are imagined and enacted, become central to the construction of self" (Desforges 2000: 929). As with the tourists in a study by Matless (1995), the sense of a moral self was bound by practices described as "the art of right living". NGO study tours were morally distinct because of the tourists' motivations that, in turn, assisted in self-definition. Secondly, these tours led to a form of moral connection with a specific form of development predicated on doing the 'right' thing. How this was defined is, of course, problematic; but if rights-based development was about preservation of dignity, then so too was tourism focused on well-being and fundamentally humanistic responses to hardship. This solidarity approach could be considered rights-based tourism, with an underlying sense of moral commonality.

Throughout all this, the participants particularly wanted to meet people who shared their interests, values and morals. This resonated with the concept of normative communitas in the personal transition literature. Aspects of Turner's (1969) notion of communitas are germane to the morale or harmony of tour groups and are particularly useful in understanding the role of peer groups in shaping the experiences of tourist strangers, as Nash (1996: 50) points out. In my research, the tours produced a form of normative communitas amongst the participants when barriers between them broke down and an unmediated connection was formed. Participants demonstrated the importance of peers: "going in a group of like-minded people gives an opportunity to share reactions and knowledge" (Ingaberg – Oxfam tour) and "*OCAA* tours are usually well organised with like-minded people ... I don't particularly want to be a stereotype tourist and visit the stereotype destinations" (James – Oxfam tour). Such statements were also consistent with the idea that personal benefits such as friendships, a sense of belonging and social interaction with like-minded people can motivate tourists' participation (Crompton 1979, Krippendorf 1987). They developed a sense of place and meaning by taking part in travel experiences with others interested in learning about development in Cuba. They saw themselves and other participants on NGO study tours as 'moral' and 'educated'. "The feeling of belonging to a community of tourists, people who identify themselves around their experiences, is an important part of [their] agenda" (Desforges 2000: 937). This resonated strongly with the way the NGOs promoted travel experiences as a unique opportunity to meet like-minded individuals, which suggests that connections were being made that might in turn reinforce new social movement participation and support for development campaigns.

NGOs and Tourism

Understanding what participants saw as the main function of the NGOs in providing a tourism program sheds light on the development outcomes tourists might expect through their experiences. There were various ideas about the role of the NGO in tourism, which included giving aid. As one US tourist pointed out:

> I see Global Exchange's function as a means of introducing people to Cuba, of acquainting them with the local needs, and encouraging them to help bring about normal relationship and dialogue between the two countries. Until then, Global Exchange, by sensitising people to Cuba's severe economic problems encourages individuals to bring badly needed educational materials, medical supplies and other goods and helps foster good will and understanding between people. (Valmia – Cuba at the Crossroads tour)

Creating 'culturally responsible' tourism was another function of the NGOs that participants cited:

> to ensure that tourism directly benefits the local people; to enable a cultural interchange between people with similar philosophies in different countries and to give travellers access to local people, organisations, grassroots experiences of life in a particular country. (Gertrude – Oxfam tour)

Some tourists saw the role of NGOs in providing tours as a way to facilitate meeting local Cuban people:

> because of their [NGOs'] contacts, their ability to get us face-to-face meetings with people we might not otherwise have met, and to see inside views of schools, doctor surgeries and a variety of community organisations, which were mostly fascinating. (Henrietta – Cuba at the Crossroads)

Mostly, people saw the primary role of NGOs in tourism as that of educating tourists in a way that promoted understanding of Cuba and development and made development aid tangible by arranging visits to projects and meeting with local grassroots organisations. Amelie communicated this perception concisely: "getting the message out to the broader community of their role in providing assistance to developing countries and how that assistance manifests in the various groups and organisations" (Oxfam tour). Tours of this nature appealed to people who wanted to participate in purposeful, educational travel: "Global Exchange stands out by offering educational, intellectually substantive tours to regions often in turmoil or targets of coercion or interference that are poorly understood in the US" (Kingston – Cuba at the Crossroads tour). Such perceptions indicated that people were seeking experiential learning opportunities geared to a sense of active engagement through travel and that NGOs were positioned to provide such opportunities:

> I see the main function of OCAAT as a method to educate tourists in a way that is not typical and in a way that tries to be culturally sensitive. Few tourists will have the opportunity to visit a school, have a tour of a healthcare system, visit farms, visit places and people that are important to the everyday life of a Cuban. (Ruby – Oxfam tour)

These statements suggested that NGOs provided opportunities that created solidarity as an element of development. The creation of solidarity through the establishment of networks was an outcome with major implications beyond just the tour itself. Inquiring into what people understood as the role of NGOs in tourism highlighted that integrating tourism with development and human rights organisations was not just about material exchange but also about far less tangible elements: emotional, educational, and affective. These exchanges led to a more positive kind of support not just about giving money, but rather sharing life histories and knowledge.

As mentioned, most participants see the main function of NGOs as facilitating exchange of knowledge and ideas and through this process establishing a strong sense of solidarity with the Cuban people. For example, Oscar described the main function of *GERT's* Sustainable Agriculture tour as providing "a brief window of opportunity to see how others live, to have soulful connection with them, to experience that people are people everywhere, and that the power of humanity is greater than any other" and Francesca saw the main function of her *OCAAT* tour was "to foster awareness and understanding among Australians of the problems of developing countries". Many participants felt that the simple act of "exchanging of information, creation of awareness" (Josephine – Sustainable Agriculture tour) and "enabling us to meet Cubans trying to make their country work so that we can find out how they are doing it" (Tom – Oxfam tour) was a valuable experience provided by NGOs through their tourism programs. This was especially important in the case of Cuba because of America's longstanding campaign to undermine the Cuban government. Ingaberg, a participant in an Oxfam tour, expresses this role of NGOs with tours to Cuba:

> to heighten understanding and sensitivity of participants to local communities in the face of the misinformation which is promulgated by the American and other capitalist press; to make participants aware of the work and needs of the Agency so that they might participate on projects, consider being active in development activities and in funding drives.

We can see that participants perceived the role of NGOs as providing them with the knowledge and grassroots contacts to establish networks by which they could support the NGOs' campaigns and foster solidarity with their new friends in Cuba.

Amongst the participants there was a sense that, through study tours, NGOs were "promoting understanding of Cubans to the world while helping raise the quality of life for the people there" (Oliver – Oxfam tour). They felt that the tours facilitated educational and affective support through a program of meetings.

Motivations underpinning participation in NGO study tours included a search for authentic encounters, educational elements with a development focus, solidarity, and re-affirmation of identity. This gives us the material to question notions of rights-based tourism and whether this is a valid category; and, if so, what are its functional implications? Increasingly, people were motivated by a desire to spend their money in ways that would support development and conservation initiatives as part of their own moral agenda. At the same time, they sought opportunities in their travel to learn about the issues affecting other people around the world and to meet people in developing countries to establish strong connections that may last beyond the travel experience. This trend fed into broader tropes of sustainability, where people made decisions based on ethical and moral imperatives. This trend thus fits squarely within notions of rights-based development in practice. The next chapter looks closely at tourist experiences and constructions of meaning in NGO study tours in Cuba, and whether they were perceived as transformative for individual tourists and, more importantly, for development.

Chapter 6

Transformation and Agency in the Tourism Encounter

Cultural research into tourism and social development in Cuba provides a situated understanding of tourist experiences as forms of cultural exchange and transformation. It appears that through experiential learning, NGO study tours transform tourists. The anthropology of tourism perspective has often addressed transformation from one state to another. However, in order to capture the complexity of transformation, I reframe it in terms of social and cultural change. If we probe tourist experiences in NGO study tours in Cuba and, we find that the tourists' transformation involved rejection of the West and affirmation of identity. Thus personal transitions in tourism were not straightforward shifts, but were nuanced and complex. One important dimension was the ways tourists and their engagement facilitated the networking of new social movements. Accordingly, education provided the platform for subsequent action, where the notion of 'meaningful' experiences had instrumental outcomes. Exploring such experiences highlighted resonance beyond the tour itself. This is of interest for what it says about tourist identities and necessarily so if particular styles of tourism make people feel morally superior. To understand this broader picture, we need to consider levels of agency that go beyond self-interest.

Social Movements and Tourism

For the purposes of this book, the term social movement "covers various forms of collective action aimed at social reorganisation. In general, social movements are not highly institutionalised, but arise from spontaneous social protest directed at specific or widespread grievances" (Abercrombe, Hill and Turner 1994: 389). "Social movements are defined precisely in terms of what they supposedly bring about: new forms of politics and sociality" (Escobar 2005: 345).

GERT and *OCAAT* were agents of network building by virtue of the intense social experiences they provided tourists. Their tours could serve as catalysts for increased social movement participation. As facilitators of new social movements, these NGOs hoped to harness the power and influence of educational, development-oriented tourism for their ability to facilitate enthusiasm and mobilisation efforts.

Through the 1980s, social movements established themselves through NGOs moving across the boundaries of nation states, producing transnational avenues of financial and political support and uniting people globally on issues such as environment, human rights, and refugees (Tsing 2000). Indeed, many post-development scholars have embraced new social movements as a means of directing alternatives to conventional development (c.f. Esteva 1988; Escobar 1995; Rahnema 1997; Sachs 1992; Illich 2001; Hall 2002). New social movements are distinguished by an expressive politics and their resistances to the 'developed society' model as the institutions of the welfare state have receded. They challenge the declining legitimacy of development in its national and global incarnations, in an effort to internationalise the realities of the indigenous actors of resistance (c.f. Guha 1983, 1988, 1996; McMichael 2004; Parajuli 2001; Spivak 1988; 1996).

As the data from Cuba reveals, new tourism niches forecast an increasing interest between globalised social movements, tourist engagement in social change, and endogenous development. Identifying these conjunctions entails exploring beyond new social movement membership (Barkan, Cohn, Whitaker 1995). Belonging to an organisation can manifest in many distinctive ways, as we have seen with the tour participants in Cuba. They demonstrated a high involvement in new social movement participation. Their support of such movements, through charities, development agencies and conservation groups, indicated their commitment to and involvement with global development issues. One member might make regular financial contributions and receive newsletters, while another participated in activism and campaigning. Additionally, as we saw in Chapter 5, many participants were motivated to join an NGO study tour as a means to financially support projects and facilitate connections with other like-minded tourists and local Cuban people.

Knoke (1988) told us that, in analysing social movements, it is necessary to include external and internal forms of participation. Typically, studies of social movement participation analyse internal forms such as volunteering in the administrative procedures, voting or running in elections, and the provision of resources. But I argue that it is the external participation that can have far-reaching effects in their mobilisation against the US dominated world system. External forms of participation include lobbying politicians, attending rallies, and, as I argue, participating in NGO study tours. These tours were a form of external participation of new social movements that evidenced a "movement of movements" (Klein 2004: 220) precisely because of their international reach, because the people who participated in them represented a mosaic of groups and campaigns that make up new social movements, and because of the networks created during the tours.

The creation of networks is an important aspect of new social movements, and this was a key characteristic of NGO study tours in Cuba.

> New social movements are rhizomic (assuming diverse forms, establishing unexpected connections, adopting flexible structures, moving in various dimensions – the family, the neighbourhood, the region) (see Deleuze and Guattari 1988). They are fluid and emergent, not fixed states, structures, and programs. (Escobar 2005: 347)

Individuals and organisations linked together through one or more social relationships become networks, which are an important element of new social movement participation and potentially reinforce activism support (c.f. Klandermans and Oegema 1987; McAdam and Rucht 1993; Knoke 1988; McGehee 2002). For example, our ties to family and friends or work colleagues can influence our worldview and our support for political and social issues. Pfaff's analysis of the revolution in Eastern Europe during the late 1980s supports the significance of the role of informal networks to social movement participation (1995). These informal ties that form our social ties are important reference points that are vital to network development, participation and commitment to social movements (Lichterman 1996: 24). This was fundamental to the success of NGO study tours in mobilising support, because the people who participated in each tour had their own social networks, to which they disseminated information about their experiences in Cuba.

The ability to make a commitment emanates from important and valued connections among individuals (Boyte 1980), like the connections forged on NGO study tours when people felt a sense of communitas with other participants sharing the same experiences. These informal connections attested that networks can be a vital source to inspire participation. Again, this corresponded with the goals of *GERT* and *OCAAT*, as they hoped that participation in their tours would lead to the dissemination of information about the NGOs' work and social development in Cuba in general, thereby encouraging others to get involved. As part of the mosaic of global new social movements, *GERT* and *OCAAT* utilised tourism as a means of educating tourists to become more active in campaigning on issues of international concern.

Likewise, McGehee's (2002) study of Earthwatch expeditions found that participation encouraged social movement engagement. Her research indicated, as does mine, the effect of networks on individuals – developed during intense educational tourism experiences – to increase participants' motivation and intention for future activism. What follows is an in-depth exploration of the experiences people had on NGO study tours that led to complex transformations and affirmations of identity politics. It was from within this fusion of processes that networks evolved to produce forms of political solidarity.

Creating Solidarity

Some time after co-ordinating my last study tour in Cuba, I received a long email from one of the tour participants, William. He was writing to update me on the past year and a half. He told me that on his return from Cuba he had joined the Australia Cuba Friendship Society, largely due to the persistent efforts of Georgia, one of the other tour participants who had also become a member. He had then worked his way through the ranks to become President, and informed me that another tour participant, Dorothy, had joined and become the Secretary. He wrote, "the Cuban odyssey was the best trip I have ever been on and I have decided to return with the Southern Cross Brigade in late December. As I'm returning to Cuba you may be able to advise me on new issues and people to meet with while I am over there; last week we met with the Cuban Consulate".

William's experience exemplified the best possible outcome that NGOs and Cuban grassroots organisations anticipated for the study tours – a strong commitment to Cuba's ongoing development. This was one example of post-tour solidarity; I cite others throughout this chapter. In this example, three people who had met on a NGO study tour in Cuba subsequently joined the Brisbane branch of the Australia Cuba Friendship Society actively participating in political solidarity efforts towards Cuba. William's email reinforced the opinions of people I interviewed and had conversations with about the personal benefits of NGO study tours, demonstrating that people consider these tours to have enduring implications:

> The advantage of this travel is a view of the sociopolitical system and life as it is lived by the people, warts and all! (Stan)

> The benefits of these tours are the direct contact with local culture, local people, and authoritative figures and experts. Other tourists wouldn't get these opportunities. (Wilhemena)

> You can learn a lot and gain access to a lot. It was informative and put me in touch with a good group of fellow travellers. (Sebastian)

> It was a different way of being a tourist and it made it possible to see more culture rather than just what was possible to find on one's own. And you get to travel with other people who are interested in learning about the same things. (Ruby)

> Access to local people, learning about the country, travelling in a way that supports local interests, not multinational corporations. (Gertrude)

> Access to organisations, people, and smart interesting group members who ask questions I might not have thought of. (Henrietta)

If we are to effectively engage the intersection of tourism and development, then we need to know what these opinions tell us – and why they matter. Firstly, they highlighted what participants saw as the main personal benefits of this form of tourism: access to the supposedly authentic 'backstage' of the local community, like-minded fellow travellers, and an accompanying educational component. All of these outcomes correlated with motivations for participation in this form of tourism in the first place. Secondly, they resonated with arguments about tourism experience and personal transformation. Some theorists have argued that tourism can offer a chance for learning and subsequently produce a personal transition through a sense of renewal or re-creation (Crompton 1979; Krippendorf 1987; Weiler 1991). Others have argued that tourists do not experience personal transformation due to the limits of their interaction: their trips are typically of short duration and thus provide little opportunity to interact with locals, and because tourists typically do not speak the local language (Bruner 1991; Laxson 1991). Many studies have examined tourist experiences in terms of the impact on hosts, and more recently, tourists (Cohen 1974; Cohen 1979; Dann 1999; MacCannell 1976; Masberg and Silverman 1996; Urry 1990).

The NGOs in this study explicitly advertised their tours as providing opportunities for transformative experiences. Nash (1996), however, pointed out that there has been little research on the persistence of attitude change due to tourism, and that there is a need for more research into whether tourism has significant and enduring consequences for the tourist and their society. Thus, while changes to the host have been long examined, changes to the tourist have received much less attention.

NGO study tour participants' experience of Cuban culture is very different from mainstream tourism precisely because of the nature of the interaction with local people. Their experiences include entering the perceived 'backstage' of Cuban culture and ways of life because they meet with grassroots organisations and discuss local development problems. The participants I talked with consider such interactions in positive terms. This clearly shows that "the experiences of tourists can be enhanced through the provision of opportunities to interact with local people and experience different cultural settings" (Ross and Wall 1999b: 677). These findings therefore substantiate claims that 'intense' rather than 'superficial' social interactions between cultures are associated with positive attitudes (Amir and Ben-Ari 1985; Ahmed and Krohn, et al. 1994; Pizam et al. 2000). The people I questioned emphatically talked about the educational aspects of their experiences.

Historical activities such as pilgrimages give us a foundation for understanding the subjective experiences on development-oriented tours. NGO study tours offered tourists an intense learning experience – people returned home with a more comprehensive understanding of development. Moreover, these NGOs hoped that participation in their study tours might encourage people to encourage others to support development initiatives once back home.

We now need to go further and examine the specific experiences of tourists participating in NGO study tours, in an effort to gain insights into their transformative qualities and outcomes for tourists and development practices. This chapter does not focus on sociocultural, environmental or economic impacts encountered by local communities in Cuba as a result of development projects and the study tours that visit them. Nor does it examine the extent to which the NGO projects have been successful in their endeavours on that level. Rather, the arena of scrutiny is the tours themselves, focusing on an ethnographic account of tourist experiences. My intention is to bring critical attention to convergences between development and tourism by examining the tourists and how they interpret and understand development issues. From this I make some claims as to the contributions of this form of tourism to development in Cuba.

Talking to tourists about their experiences reinforced my impressions of the viability of this form of tourism. For example Oscar, an American horticulturist, emphasised the importance of expanding these initiatives. In his opinion:

> getting the general public on this mode of visiting another country is critical. I think many people would shy away from Global Exchange's 'radical activist' side. Those who are less likely to sign up for a cultural awakening are certainly not going to if the tour company is also perceived as 'radical' or 'threatening'. The challenge is to maintain the activist aspect without coming across as anti anyone.

Oscar's opinion about the importance of this form of tourism shed some light on why it is worth investigating tourist encounters with development. That is, once development begins to utilise tourism, it unleashes a potentially useful but ultimately uncontrollable force. To some degree, tourism provides uncertain, complex, and unpredictable sets of variables making it important to provide an analysis of specific tourist experiences.

Here I address the question of whether tour participants perceived their 'culture contact' as more meaningful and productive of further engagement than would be the case in mass leisure tourism. For example, did development-oriented tours lead to exchanges beyond immediate financial support that effect the social and cultural environment of local people in positive ways? Hodge (2004: 17) argued that, through the sharing of ideas with both local people and with people back home, volunteer participants could promote a greater understanding of some of the issues regarding systems of aid and development, thereby increasing their contributions to development initiatives.

This leads us to further a central tenet of this book: if tourism and development are becoming entwined in various ways, then more attention should be paid to the tourist experience in specific ways. The question more importantly becomes: does a meaningful engagement, which embodies a particularly powerful experience for tourists, lead to a process whereby tourists become actual agents of development? Over ninety per cent of participants stated that they did not feel their experiences were transformative in terms of their attitudes and behaviours, but rather, their experiences had led to a more nuanced understanding of Cuba and development. Thus transformation occurred as experiential learning where aspects of identity were enhanced. For most, participation in NGO study tours was a reaffirmation of certain identity characteristics rather than a transformative experience in its own right. However, if their experiences led to a more informed understanding of Cuba and the role of development processes then we could argue that this indeed facilitates a transformation from tour participant to agent of development, which occurs both within their exchanges with local people and on their return home as they disseminate information about Cuba and its social development. Therefore it is necessary for us to look at the terms of transformation and the ways that through experiential learning tourists acquire agency.

The point in addressing tourist experience is to gain an insight into what ways tour participants perceive their culture contact as meaningful. Referring to volunteering experiences, Hodge tells us that a principal achievement is "the cultural exchanges that take place between volunteers and locals, and the heightened level of understanding which impacts on both cultures. Such exchanges have become critical in the current global-political climate – their value cannot be underestimated" (2004: 15). This perspective applies to NGO study tours too, because, they perpetuated links between countries by creating networks that fostered solidarity. In these ways, participants considered that their culture contact was more meaningful than any such contacts experienced through mainstream forms of tourism.

As the responses from people I questioned indicated, one benefit of NGOs operating study tours was the promotion by participants of the NGOs' philosophies and objectives, as it was important for most tourists to share their tourist experiences back home, typically through photography, video and travel tales. This was also important to the Cuban organisations involved with the study tours, as they saw the tours as a means of promoting solidarity. Ninety eight per cent of participants stated that they would *continue* to be involved with charity and/or activism, that they learnt much about Cuba, and that they had emphatically positive encounters. Despite a lack of consciously recognised subjective change, I argue that indeed a transformation is taking place because there was an increased commitment specifically to supporting Cuba by many participants. What the following discussion tells us is that NGO tour participants were politicised people and that many of them, on returning home, wrote letters to their politicians, wrote articles in their local newspapers and community newsletters, gave presentations, talked to friends, family and colleagues disseminating what they had learnt about Cuba, and they even joined their national Cuba Friendship Society. For example:

> I believe the Cuba at the Crossroads tour introduced us to important movements in Cuban culture and politics. The tour made us aware of the difficulties facing the Cuban people and government and sent us home determined to help in their struggle. The Global Exchange tours are helping to educate Americans about Cuba. (Henrietta)

It is in these ways that we can conceive of NGO study tourists as agents of development, even if only to a minor degree. A shift took place when tourists returned home and boosted either the development efforts of the NGO they travelled with or the solidarity efforts with Cuba. Prior to their Cuban tour, they were not directly involved in supporting Cuba in these ways, but on their return they were relatively active in their support for Cuba, even if they were already involved in other charitable efforts. Thus, while many participants stated that they did not experience a personal transformation because they were already active supporters of development causes, it can in fact be seen they become more involved in supporting development and solidarity efforts of Cuba specifically (over ninety per cent of participants).

This suggests that this form of tourism achieved an exchange beyond just a material and immediate social level. The elements most participants cited as significantly impacting on their experiences and as highlights of their tours were the seminars and project visits, the time they had to meet local people and make deep connections, and the opportunities to undergo these experiences with like-minded travellers. "The exchange of ideas and thoughts and friendships with local people and with each other are extremely satisfying to the tourists and I think the local people too" (Innes – Oxfam tour). Thus it was the educational, the intellectual, and the affective exchanges with other participants and, with local people that were significant. Tourists arrived with aspirations to learn about development in Cuba from local people and they left with knowledge about development and a relative commitment to Cuban solidarity. The seminars and project visits represented opportunities to engage with local people on an intellectual and emotional level.

Exchanges occurring in a tourism development context arguably improved the social and cultural environment for the local people because they were empowered to create global networks through these touristic exchanges that celebrate their cultural, agricultural and political diversity compared to the US model. The impact of such experiences could be seen to transform tourists into nascent agents of rights-based development because of the overall sense of fleeting well-being for local people that such exchanges facilitated and the solidarity links they cultivated.

It has not been my intention to analyse the host experience, such as the well-being and dignity of local Cuban people as a result of these tours. But in my endeavour to investigate this form of tourism, I can make some claims for its contribution to rights-based development. What we can see taking place is a form of development that is more than just about structural adjustment leading to economic development which is so prevalent throughout Latin America. It is a form of rights-based tourism where local Cuban people who are actively engaged in the development process also create networks with international tourists through NGOs in an effort to foster solidarity. Creating solidarity through networks is linked to the capacity of tourism to bring about subjective changes in the conditions of Cuban people. Arguably, this form of tourism promotes increased self-respect and self-realisation of local people and tourists alike through a program of educational and affective exchanges that enhance empowerment and independent agency. These are tentative propositions. But the weight of these social contacts cannot be dismissed and thus what follows is an exposition of the transformative nature of NGO study tours through experiential learning that leads to these productive outcomes.

Transformative Learning: Deepening and Enriching Travel Experiences

While there is still considerable debate as to whether different styles of tourism produce enduring personal transformations,[1] a critical question that underpins research into the conjunction of tourism and development is how tourists' experiences (whether transformative or not) are integral to the macro factors that influence tourism. For example, has this form of tourism emerged out of larger tropes of development, or is it being driven by the tourism decisions of a more discerning, Western, new middle class? If aspects of development are intersecting with tourism, then its ongoing sustainability will depend greatly on tourists' experiences. Consequently, development specialists need to consider what people gain from their tours.

Notions of transformation are complex and cannot be understood merely as a direct transition from one state to another. For the majority of NGO tour participants in Cuba the tours *reinforced* existing attitudes and behaviour precisely because most claimed to already contribute to development efforts through financial support, activism, fundraising, and so on. Furthermore, most of them considered participating in a NGO study tour as part of a broader inclination to support development work. The tourists revealed in their conversations that they made greater commitments. Transformation was therefore not a black and white process. This can be shown by a number of practices, some not always self-consciously evident to the tourist.

These tourists were not just interested in meeting the exotic 'Other' as objects to be viewed, but, rather, they were engaged in what they perceived to be meaningful contact with local people in order to exchange ideas and information about development issues. They could disseminate this information on their return home in an act of solidarity and contribute to the development efforts of the organising NGO. The overall effect of short-term experiences on attitudes was mostly different for younger participants than it was for the older participants. Younger members of the tour groups explicitly stated the tours did have a transformative effect on them. Presumably this is due to the tour being their first such experience. The younger participants said that the tour had affected a personal change and inspired a commitment to development. All the younger respondents indicated that their tour did change their previous ideas *and* affected their prior knowledge of Cuba. Wilhemena discovered through the seminars and projects many contradictions in Cuban society:

---

1   For examples of transformative tourist experiences see Orams' (1997) study on ecotourists at the Tangalooma Dolphin Feeding Program in Australia and Tisdell and Wilson's (2001) study of ecotourists at Mon Repos Beach in Australia. The informants in these studies claimed to be more environmentally responsible in their subsequent behaviour.

urban agriculture seems to be communism at its best. In that sense I grew to appreciate how communism can care and provide for its people (for example the agriculture project we visited that was growing for the elderly). In a similar light, I understood how it was the introduction of capitalistic ways (like selling produce at the markets) that was spurring the economy into a state of recovery. So many contradictions! The tour definitely changed my prior understanding of Cuba. I didn't know much of Cuba except the anti-Castro propaganda the US government spouts.

Oliver felt that since his tour he had "a much broader base of knowledge about the country especially compared to the average American" and for Ruby "the trip has affected me tremendously. I understand Cuba much better than I ever thought I would. I really knew little about it before I went. I think the trip was so effective because I was not just a tourist in Cuba; I was being educated along the way". The tours offered people enriched experiences of Cuba through the lens of development. This could be because the younger participants were all US citizens. Nationality emerges as an important aspect because as young Americans they had never really challenged the anti-Cuba rhetoric promulgated by the US government and media. Thus the tour had a strong impact and they felt that they had learnt a lot while on their study tour. Ruben exclaimed "it is such a great experience because it affects you and you have to change". This could be because they did not have the lifetime experiences of older participants, whose lifestyles are more in accordance with this type of tour, so that it did not emerge as a unique experience for older participants. For the younger participants, the tour was unique and had such an impact that they felt compelled to transform the way they travelled in the future and to become more active in supporting development initiatives once back home.

Several younger participants indicated that they believed they had been changed but that they might not know exactly how for some time. Ruby was "not quite sure what the lasting effect will be on me, but I know it has caused change in my everyday life. I don't think a day has gone by where I didn't think about my time in Cuba, making it seem almost like a dream". For others in the younger category, the personal transformation was quite clear and their statements illustrated the tour's power to affect participation in new social movements through supporting development. Sebastian said: "I want to be involved in development and community development issues and become more of an activist now"; and, for Wilhemena:

the lasting effect on me is that I will be engaged in Cuban politics and human rights from now on. I think Cuba/US relations are a fascinating story and I think that from now on Cuba news/culture/politics will be a hobby of mine. The trip has inspired me to grow my own vegetables – this is certainly a lifestyle change. I am inspired to spread the word about urban agriculture in Cuba and to encourage people to visit. I recently held a slide show on Cuba and will present another one next month at the University of Michigan. I also hope to write an article to further disseminate information of Cuba's unique model of relieving food crises through urban agriculture.

The personal transformation expressed by Oliver resulted from seeing how materially poor Cuban people were but how happy they appeared:

> I become more and more humble about materialism. I have learned to be happy with what I have, because it is still much more than many people in the world. I try not to let small things get to me anymore, because now it just seems stupid to stress over minor inconveniences in life. I have food, clean water, healthcare, and a roof over my head. What more could I really ask for?

Oliver's idealism reflected images of utopian community and simplicity that many people in the West feel alienated from, as evidenced by the increasing mobilisation of new social movements. The contrast between the 'transformative' experiences of the American youth visiting Cuba and the desires of many of my Cuban friends who saw America as a utopian paradise of wealth was striking. When not on tour, I often went with Cuban friends to outdoor hip-hop concerts, which were one of the only domains where young Cuban artists could voice their political opinions. Themes they sang on were often the contrasts between Cuban society and what they imagined American society to be like. Even the style of their hip-hop emulated American popular culture that they sourced from music CDs circulating on the black market or sent to them from their relatives in America. Many younger Cubans do not express the same revolutionary spirit as the older generations and paradoxically their imagined America is a kind of utopia.

In contrast to the explicit transformations expressed by younger tourists, there was a remarkable difference in responses from the older tour participants, which they interpreted as a result of their existing engagement and awareness of development issues and that they lead fairly simple lifestyles. This is interesting for what it tells us about the transformation thesis of much tourist experience literature. It suggests, as we shall see, that tourist experiences involve complex negotiations that cultivate cultural change rather than an utter transformation *per se* for the majority of tourists participating in development oriented study tours.

A number of people made a clear distinction stating that "my previous ideas were not changed so much as they were expanded and my understanding deepened" (Kingston) and "not so much change as fill in details and intensify" (Henrietta). Gertrude's "prior knowledge and understanding has deepened" about Cuba as a result of her tour; likewise, for Valmia: "my tour has certainly increased my understanding and appreciation of the hardships Cuba is having as a result of the embargo and natural disasters". Participants used positive words and phrases to emphasise how meaningful the experience was: 'very'; 'extremely'; 'incredibly'; 'invaluable'; "the overall experience has been meaningful because it has given an insight into a country of which I knew very little" (Innes); "the overall experience has been extremely meaningful for me now that I can put things into perspective and in context in relation to anything Cuban" (James); "one of the best trips ever and intellectually started me thinking and acting in a different direction" (Ingaberg).

We can conclude from this passionate affirmation and the emotional terms used by tourists to recollect their experiences that the new knowledge gained and their experiences greatly moved them. As we will see through participant accounts, horizons were extended for some tourists, the individual dignity of local Cuban people they met with was valued, and the trust and solidarity that the groups created through sharing histories, difficulties or goals *potentially* fed into giving local Cuban people some degree of support in their attempts to reinvent themselves as more active subjects in their communities. For the tourists, the high levels of interaction with other tour participants, local officials, and local people in the community facilitated the establishment of network ties that are so important to the success of new social movements (McGehee 2002).

Significantly, all participants acknowledged an increase in their knowledge, whether it was transformative or building on their existing knowledge, relating to analyses that depict 'learning' as part of the new moral tourism experience. Here we see that participants on NGO study tours considered that they learned about key development and sociocultural/political issues. This reinforces the values and characteristics associated with new moral tourism where tour participants in NGO study tours to Cuba demonstrated a sense of contributing something interpreted as 'good' and gained a sense of moral superiority from the positive outcomes associated with their tours. For example, it is implicit in Josephine's statement that her experiences afforded her a moral point of view, because they involved 'backstage' access to Cuban culture; thus, she felt she was more aware of Cuban realities than other tourists: "the tour presented another view on Cuban culture and Cuban issues allowing for me to see and decide the situation. I'm aware now of Cuban realities from Cuban perspectives".

Many participants' experiences in Cuba reinforced their commitment to supporting causes both at home and abroad. For example, Ingaberg experienced "an awakened interest in gardening and a reinforced notion that organic agriculture is something that the West could embrace more fully. As a whole the trip solidified ideas that were brewing already in my head". Likewise, Valmia's tour reinforced her "will to continue supporting as many environmental, family planning, and civil rights organisations that I can afford. I've always been an activist and will continue to contribute and disseminate pertinent information and lobby congressional members". Josephine confirmed that "the tour has revived my commitment to my conservation approach to living and being vocal about issues of social justice and environment … I have since written articles in our community newsletter and sent emails to staff and friends about my learning experiences in Cuba". For Stan the tour "has served to support my prior resolve to lead a relatively simple, environmentally responsible lifestyle, sensitive to the plight of others around the globe and the social injustices they suffer". James said: "the tour has confirmed that I do need to continue being vigilant about not only Cuban, but Australian matters such as detention of asylum seekers, Aboriginal reconciliation, racism, unemployment, foreign affairs. Everyday people have to have a voice". These statements highlighted the politicised commitment of tour participants and the opportunities provided by the tours to exchange information about networks and to develop ties.

These statements also resonated with themes of idealism regarding community and simplicity held up as alternative to the consumption and alienation in the West. As discussed in reference to the new middle classes and new tourism literature, we have seen a shift in how people live in the West. In Western societies, more people are becoming involved in new social movements. This shift is a direct response to an increasing alienation from a society in which people are in search of more meaningful experiences elsewhere, as MacCannell (1976) suggests.

Hollander wrote that "the utopian susceptibilities of contemporary Western [tourists] are part of a long-standing tradition" (1981: 29). Many of the tourists involved in NGO study tours demonstrated very strong similarities to Western intellectuals who travelled to the Soviet Union during the 1930s. For example, Tom, who joined an *OCAAT* tour was typical of the middle classes and their lifestyle choices. He worked in renewable energy design, his wife grew organic vegetables, they had travelled to former Soviet countries and they were active members of the Nottingham Refugee Forum. Francesca also joined an *OCAAT* tour; she actively supported charities and human rights groups with fundraising, lobbying, organising and public speaking along with paying financial contributions and memberships to various welfare and conservation agencies. She had also participated in a number of NGO study tours and travelled independently. Gertrude was another example; working in special education, participating in NGO study tours often and supporting charities. Josephine joined the sustainable agriculture tour with *GERT*. Talking with her, it became apparent that the style of tour accorded with her sense of identity and other lifestyle choices; she was already heavily involved in sustainable agriculture at home and at work. She wanted to learn more in order to incorporate the Cuban model of urban and organic agriculture into her projects.

The tours facilitated an intensely emotional interaction that provided a sense of connection with the way of life in Cuba beyond what are depicted as superficial encounters. While we cannot generalize that all NGO tourists going to Cuba had these experiences, my findings suggest that the majority share these positive responses. The experiences tourists had engaged them at a particular level of action that enabled them to feel they were doing something positive. Although only the younger participants considered they had experienced a personal transition, there was also a level of change in the other participants' behaviour in the form of increased solidarity efforts on their return home. At the very least, new relationships were established, allowing the creation of networks between the tourists and local people they met, and also among the tourists themselves.

Thus, tourism and development created new kinds of solidarity and transnational connections. These connections were didactic and provided productive experiences with local Cuban people. McAdam's and Rucht's model of information exchange within social movements (1993) demonstrated that the dissemination of ideas and activities on a global scale occurred through such networks. To understand the moral imperative driving the conjunction between new social movements, development, and tourist concern with social change, let's now take a closer look at the tourist experiences that led to the transformations of which I have been speaking.

Travel as Experiential Learning

It is clear the tourists gained substantial knowledge about Cuba and its distinct form of social development in the course of the tours. In fact, many participants said that they were "the most important parts of the trip for me" (Kingston) or more specifically the seminars and project visits were "the most important facet of the tour in providing vital background material and the opportunity to question grassroots personnel closely which assisted in putting past and future tour events into perspective" (James). "Without these visits and seminars it might otherwise have been just another holiday tour, because we had opportunities to really talk with local people about important issues" (Amelie). Ingaberg pointed out that the meetings with key government, development and grassroots figures and visits to community projects were "the highlight and something money wouldn't have been able to buy if travelling alone". Thus, for the tourists, authenticity was achieved by viewing development practices at work in community projects and discussing issues with local peoples.

The tourists not only learned but also interacted in ways that confirmed certain points of view, such as the authenticity of their experiences, which were qualified specifically in relation to 'the local people'. Tour participants viewed the development-focused meetings and project visits as the key point of difference from mainstream tourism precisely because they provided opportunities to meet local people in order to learn from grassroots perspectives. For example, Tom, from an Oxfam, tour said: "there is no way we would meet these people without coming on this type of tour" and Gertrude felt: "these aspects were very important to me as they are very much the reason for my choosing this type of travel. I would not have had access to meetings with local people, which for me brought the trip alive". Learning from the grassroots perspective was seen as a fundamentally distinctive aspect of the meetings and project visits. Stan felt that the "seminars with key figures brought out information at the management level whilst visits to projects gave a hands on at the coal face perspective" and Oliver told me "they provided some very interesting information that allowed us to put together our own view of Cuba as a country". The local point of view gave Ruby the opportunity to gain some insight into Cuban lives: "it was very important for me to visit the sites we did and learn from as many different perspectives as possible ... these places offered some great information on the lives of Cuban people as well as Cuban thought".

These comments suggest that participants felt they gained entry into the lives of Cuban people and that it was the NGO that provided this access to the 'authentic backstage' region. The importance of the seminars and project visits can be understood in terms of how participants viewed the role of the NGO. For example: "the Oxfam Cuba tour helped us talk to a lot of people we would not otherwise have met" (Tom); "Global Exchange got us in contact with local people" (Bain); and "I chose the Oxfam tour because I felt it would possibly offer more contact with and insight about the country, its people and their culture and politics" (Amelie).

However, at times the busy schedule of meetings tested the level of tolerance for such experiences. Participants spoke of being occasionally overwhelmed by the volume of information they received each day and said that they would have liked debriefing sessions in the evenings in order to process it in a more meaningful way: "Some of the days were quite heavy going, but it was worth the effort. Many times I thought of questions after the seminars; maybe we needed a daily debriefing session for these" (Tom). Amelie agreed:

> on days when we had two seminars or visits in one day it was difficult to digest, as it is all foreign to us and requires a lot of thought to get everything into some sort of order, to be able to analyse what we've seen and discussed on that day; nowhere near enough debriefing. Debriefing after each or every second visit would have been immensely helpful in clarifying points of issue.

Rather than demanding more leisure time, many tour participants wanted the opportunity to consolidate what they had learnt each day in order to maximise the educational benefit of their tour. This resonated with the education theme identified as an important motivator for participation in NGO study tours. Ingaberg valued the idea of formal debriefing sessions for the opportunity to discuss her thoughts and hear those of her fellow group members: "I would have welcomed some formal debriefing and discussions as a group in the evenings for lots of good potential exchange". This related to the importance of peers on such tours and to the concept of "normative communitas" where people communicate with other people involved in the same process at the same time (c.f. Turner 1969).

Overhearing and participating in conversations on bus journeys through the provinces and at meal times, I observed that participants benefited from their rich and sometimes intense discussions and debates with one another about Cuban development issues. Indeed, the group members contributed to the overall experience for many of the participants because they valued each other's ideas and thoughts about their shared experiences (c.f. Turner 1969). This reinforced Boyte's (1980) findings that social movement participation arises from important and valued connections among individuals.

Although the itinerary of meetings and project visits had been confirmed months in advance of each tour, cancellations were not an uncommon occurrence. This occurred at least once on each tour that I co-ordinated or participated in. Conversations about the cancelled meetings revealed how important the educational component of the study tours was. The late cancellation of meetings was usually a great disappointment for many participants, once more highlighting the particular type of 'authenticity' desired: 'local' contact on the tourists' own terms. The cancellation of one seminar or project visit could mean missing out on learning about their specialised interest from local people. For example, Stan said, "unfortunately one important project visit didn't eventuate, visits to properties of ANAP members" and Henrietta from another tour commented: "I do wish we'd been able to visit a doctor's office and a school; both visits were cancelled last minute". Statements of this kind demonstrate the importance of local contact, the niche interests of tour participants, and that these special interests are not always met. Tour participants were made aware that there was a distinct Cuban concept of time that lacked any sense of urgency and cancellations were explained in terms of this cultural difference.

When discussing NGO study tours, we must be careful not to see all interaction as always viewed in a constructive light. Some participants complained to me about the propaganda from some institutions, feeling they were being treated as any other international tourist, which compromised their educational agenda. Mostly, participants recognised that propaganda typically stemmed from the government sponsored visits: "the party-line drivel we encountered from any official contact we had became a bit depressing" (Marcel) and Henrietta viewed these seminars "as in some degree the sort of bullshit you'd expect to find at the government sanctioned visits; the Committee for the Defence of the Revolution block party leaps to mind". While there was always an official line touted by government organisations we met, the itineraries were designed to involve meetings with both government and grassroots organisations in order to provide broad perspectives on social development. However, some participants remained sceptical, assuming that the seminars were all a façade: "One of course had to sift what we were told because a lot of it was propaganda. Our wonderful guide was a programmed man but as long as that was recognised I had no difficulty in comprehending what the issues were all about" (Innes).

Experiencing the 'local', 'real', 'authentic' was expressed as very important; yet some tour participants felt that their encounters seemed programmed and too planned. Such statements indicated a perception among some tourists that the presentation of Cuban reality was purposefully selective concurring with Hollander's (1981: 372) portrayal of the highly organised political tours to the Soviet Union. He referred to the "techniques of hospitality" used to ensure tourists gained positive impressions of the social system; and, indeed, some participants felt this technique was being employed for them.

Tourist attractions were another low point for many participants. They felt that, through NGOs, they had access to the 'backstage' of Cuban culture and were not interested in participating in the frivolity of mainstream tourist attractions. Styles of tourism are often predicated on this sense of moral and economic superiority. While rights-based development aims to diminish Western claims to moral superiority, the rise of new moral forms of tourism has a tendency to encourage an alternative version of moral superiority. Francesca "loathed the touristy cabaret at our hotel in Sancti Spiritus" and James found "the touristy set ups like drinkies at our arrival to some places so unnecessary". But, on the other hand, many people were happy to accept these aspects of their tour. Some participants thought some of the hotels were unpleasant: "the hotel we stayed at just after Las Terrazas was beautiful but unpleasantly touristy" (Oscar), and Ingaberg complained to me: "I hated being at the Club Med style resort at the dreaded Varadero, I contemplated leaving the tour at this point". When taking NGO study tours, participants did not expect to be involved in mainstream tourist activities; they expected to be utterly removed from popular tourism, as if to ensure they were not contaminated by it. But the above discussion demonstrates that the nature of tourists' experiences is complex and sometimes there are contradictions between their moral desires and their tourist inclinations.

Learning the Cuban Secret to Life

Visiting community projects, experiencing diverse dimensions of Cuban culture during free time, and socialising with other tour participants emerged in conversations as being the highlights of tour experiences. Conversations with the tourists about the highlights of their tours revealed some insights that resonated strongly with mainstream tourism imagery but at the same time revealed a sense of alienation within their own lives. Visits to community projects and agricultural cooperatives and farms in particular were important to the tourists because they represented authentic encounters and, in the case of the agricultural meetings, a chance to visit the peasant, the *guajiro*, and experience organic, manual productivity and communal life. This kind of imagery of 'the peasant' as idealised 'authentic' also resonates strongly throughout mainstream tourism. But the NGO tourists' overt engagement and educational exchanges with local people indicated a desire for connectedness with Cuban people because of the communal traits in Cuban society. Hollander (1981) reminds us that the idealist susceptibilities of Western tourists are part of a long-standing tradition.

Visiting community projects gave participants opportunities to view the various types of social development efforts taking place in Cuba. The agricultural visits were often cited as a highlight for different reasons. For Oscar, "a highlight was the farm lunches at the organic farm co-operatives and the Havana garden at Wilfredo Peres". For Wilhemena, it was "talking to the farmers and touring the market gardens". Others, like Amelie, singled out: "visiting the transformation workshop in Atarés and hearing about their ongoing projects like water, street lights, sewage, and social programs". For others, projects such as *el Comodoro* were significant: "visiting the resettlement for displaced people on the outskirts of Havana was such a great experience made especially memorable by the wonderful attitude of the persons in charge and the happiness of the children we met there" (Innes). It was a common response to accentuate positive experiences after a tour, as it is with holiday experiences in general, but it is the underlying sensibility of education and interconnection being expressed in these statements that I am particularly interested in.

Figure 18 Produce from vegetable garden is sold at local markets

Figure 19 Tour leader Emilio translates about local organic farming practices

Figure 20 Lunch at the farmhouse; organic produce

**Figure 21 Agriculture visit in Havana province focusing on organic
and sustainable crops**

Figure 22 Farmer and ox-drawn plough

Figure 23 Sugarcane in the *Valle de los Ingenios* Sancti Spiritus province

Figure 24 Local gardens grow produce for school lunches

Figure 25 Visit to a tobacco plantation; tobacco drying house

Figure 26 Drying tobacco; Pinar del Rio

Figure 27 Hanging out on *el Malecón*

Figure 28 Playing in an empty lot in *Habana Vieja*

Josephine's sentiment about experiencing Cuban culture during free time was an example of one of the substantive experiences many of the participants remarked on: "e*l Malecón* – the meeting, singing and dancing with local people who were just hanging out in the hot evenings by the sea – that was truly special". Henrietta commented that:

> there were lots of cultural highlights for me often during my own time ... a morning spent talking with some pharmacists totally out of the blue; talking with kids on e*l Malecón*; the baseball game in an empty lot in old Havana; and Cuban music everywhere!

Oscar's spontaneous drumming lesson in Santiago de Cuba:

> the private drum lesson in a man's home was a real highlight. He solicited me on the street. I declined but we continued to talk. He asked again and I agreed. He

is an unemployed percussion instructor and my connection with him was deep
... dancing at Casa de Africa was a treat too.

While such encounters were spontaneous for the tour participants because
they took place outside of the structure of the tour itinerary, they were clearly
not always unplanned for Cubans who approach tourists on the street to sell their
goods or services as a way of making US dollars on the black market. Yet it was
another level of contact welcomed by most of the participants. Often, the tour
incorporated optional leisure activities in the evenings, such as dance lessons,
which were always well attended and listed by many as a highlight, reinforcing the
nostalgic and romantic impressions of Cuban music which have gained popularity
in the West.

Socialising with other tour participants was an important element of the tour
experience. According to Nash, "the concept of communitas, as it relates to the
morale or cohesiveness of tour groups has been useful in pointing to the role of
peer groups in shaping the experience and reaction of tourist strangers" (1996: 50).
Members valued opportunities to share the ideas and impressions of fellow tour
members and their accounts reveal the importance of networks: "Being part of a
socially committed group and sharing ideas and responses to the behind the scenes
trips, which were all excellent in different ways, was such a highlight for me"
(Ingaberg). Others conveyed similar positive sentiments about spending time with
their tour group members: "I really enjoyed the company of our group" (James);
"the group itself was a highlight for me; what an interesting bunch of people"
(Henrietta); and for Ruby: "I think the relaxed nature of the trip made it better than
I was expecting, I really enjoyed the freedom at night to go out with members of
my group in Havana".

The notions of liminoid time and communitas help explain how significant the
free, unstructured time in the evenings was for participants precisely because they
spent this time with other group members and enjoyed the experiences of meeting
local people together; dancing at local hangouts like *Casa de la Musica* or *Casa
de la Trova*; hanging out on *el Malecón* drinking rum; playing guitar and singing
Cuban songs with locals. The group was important to the tourists because it was a
way of experiencing 'communal' life; their group becomes an imagined village of
connection. Again, this reinforces the idea that contemporary lifestyles produced
alienated beings in search of more enriching experiences in other cultures.

We can see a relationship emerging between participants' motivations for
participating in NGO study tours and their experiences. For instance, seeking
authentic and educational experiences was identified as an important motivator for
many tour participants and emphasised as a point of difference from mainstream
tourism. Participants could be expected to consider aspects such as 'visiting
community projects', 'meeting with Cuban organisations' and 'experiencing Cuban
culture during free time' as providing these authentic and educational experiences.
This was consistent with arguments in tourism literature that many in the new
middle classes chose holidays that were alternative, educational, small scale and

perceived as morally superior to mass tourism. Furthermore, the positive aspect of 'socialising with other participants' corresponded to the many participants' identity motivation The group gave participants a sense of community and commonality, the traits they admired in Cuban society. This last aspect provides support for claims that new tourists choose alternative forms of tourism that reflects lifestyle and reinforces aspects of identity by mixing with like-minded travellers (c.f. Mowforth and Munt 2003; Poon 1989, 1993). This form of tourism is a growing niche that potentially has positive implications for the tourist, their home society and development in general.

Cuba's Alternative Development:
"Great Crises Always Deliver Great Solutions"[2]

The combined effects of the breakdown of the Soviet Empire and the tightening of the US embargo created a "great crisis" for Cuba and, consequently, the government created a model of social development different to most throughout Latin America. Tourists learned about this specific model through attending seminars and visiting development projects. Inquiring into what is of particular interest to tourists on NGO study tours informs us about transformative learning. Discussions with people on the tours (especially conversations directly following the meetings) revealed that learning about Cuba's development issues produced tourists with more reflexive and critical interpretive awareness of the government's attempt to improve its geopolitical position through international tourism.

One aspect of Cuba's development model that had a particularly discernible impact on tourists' experiences was urban and organic agriculture. Many of the tourists had travelled there specifically to learn about this aspect of Cuba's recovery from deep economic crisis and felt impelled to implement their newfound knowledge on their return home. Cuba experienced an agricultural crisis in the 1990s that saw people go hungry because suddenly farmers had no fuel to operate their tractors, no pesticides to protect their crops, and no spare parts for their pumps. In addition, the lack of foreign exchange meant that Cuba could not import basic food. A combination of land reforms and alternative production techniques have since revolutionised Cuban agriculture and the subsequent organic conversion of the agricultural industry is a development model that is being transported to other Third World countries.

Wilhemena, an American student conducting research for her Masters course into the agricultural revolution in Cuba, told me:

> I was inspired by the country's model of urban agriculture and how it could help other developing countries ... I was glad to hear the experts acknowledge that urban agriculture was born out of necessity but that they believe that they have

---

2   Title of speech by Fidel Castro at the South African Parliament, 4 September 1998.

created a very worthwhile program that will remain in place as Cuba's economy
eventually recovers.

Oscar said: "you know something that had a particular impact on me was how
urban, organic, social agriculture is one hundred per cent possible, including with
limited financial means". For him, Cuba's organic and collective agriculture showed
a viable model to implement in feeding poorer sections of his own community
in Philadelphia. Likewise, Sebastian said: "the amount that can be done with an
unused plot of land in the city is so impressive and has got me thinking". In fact,
Sebastian indicated that "meeting the farmers stands out as the most significant
aspect of the tour, although the more theoretical sessions also do. Cuban people
seemed less interested in ownership of success for themselves than in the West;
this was really apparent with the agriculture co-operatives and their production".
What appealed to him were these more communal aspects in Cuban society that
do not prevail in capitalist societies. The tourists were interested in learning from
the *guajiros* because the rural peasantry is deeply invested in transforming Cuba;
they are symbolic of Cuba's success and in exchanges with *guajiros*, a sense of
dignity was discernible. Indeed, notions of the pastoral resonate quite powerfully
in Western tourism to Third World countries.

The Western pastoral tradition originated in the first half of the third century
BC with *Idylls*, Greek pastoral poetry composed by Theocritus and later by
Virgil in *The Eclogues* (Day-Lewis 1983: xiii). It emerged with the rise of
industrialisation and rapid growth of urban centres. The myth idealises the rural/
pastoral as a simpler, purer life and was transmitted through Western culture
first by the oral tradition of pastoral songs and then by the written translation
of Virgil's work (Short 1991: 30). Pastoral literature has typically contrasted
urban life unfavourably to rural simplicity. The myth focuses on rural life as more
wholesome and spiritually rewarding and portrays the idealised countryside as
'Arcadian' in contrast to the anonymity of the city. Western tourism evokes the
pastoral by routinely portraying Third World countries as the idealised opposite
– the rural idyll – where the countryside symbolises a refuge from the rigors of
modernity. Travel operators draw on the pastoral myth to represent Third World
cultures as romanticised, idealised and timeless, implying that they have not yet
been affected by modernity. This resonates with the NGO tourists' 'imagined'
views of community and simplicity in Cuban society. But even more prominent
was that, through the process of engaging with and learning from Cuban *guajiros*,
the tourists critiqued the West and its associated consumption and alienation as
part of the new social movements.

Certainly, it was clear that participants on NGO study tours had niche interests
connected to aspects of Cuba's social system, constructing them as politicised
tourists. But these tourists were similar to any other in some ways. Complaints
about hotel food were a reminder that they also demonstrated the expectations of
any other tourist when it came to accommodation and food: familiarity, comfort,
Western standards. They wanted a particular type of 'authentic experience', meeting

middle class comfort needs, and in these ways they remained the same as any other tourists. It might seem inconsistent that tourists who travel to Cuba to learn about development issues and claim to be 'responsible' and 'sensitive' travellers should complain about meals in the hotels but this was common. For example, one participant on an Oxfam tour commented: "the appalling food served at Hotel St Johns I put down to bad management because we saw what was available in other places such as in *paladares*, at the resort in Varadero (obviously anomalous experience there!) and lunch at the biosphere reserve was great too" (Amelie).

As we have seen, throughout the tours participants learned about the sustainable agriculture revolution and how Cuba was forced into a situation of achieving food security through sustainable measures. Participants witnessed the successful urban agriculture and farm co-operatives in the provinces. Perhaps it was for this reason that they were surprised at the poor quality of food being served in their hotels. The hotel meals represented a huge inconsistency with the farms and co-operatives visited on the tours and the well stocked fruit and vegetable markets on the streets. Sebastian agreed that "the food in hotels was bad; I expected better food" and he suggested "eating more street food would solidify the connection between what we saw in the gardens and what we put in our bellies". However, not all tour participants were of this opinion and instead complained about those participants who were dissatisfied with the hotel food: "there were some low points, the times when members of our group complained about food. Some people are apparently ignorant of conditions in Cuba and the role of an *Oxfam CAA* tour; it's not to stay at 5 star hotels!" (Francesca). This discussion highlighted the intricacy of tourists' experiences and the contradictions that sometimes arise between their backstage desires and tourist inclinations.

Many participants said learning about the effects of the embargo on the day-to-day lives of Cuban people was a theme that affected their time in Cuba. For example, Valmia said "I think the main learning point for me was the capacity of the Cubans to use their intelligence, ingenuity, and creativity to compensate for the hardships and deprivations they've been forced to live under these past twenty years". In the same sense, James "learnt about the determination of the people to be self sufficient, create their place in the world and overcome obstacles placed in their pathway and about the many opportunities for development". Innes said he "was surprised by the development of two different groups of Cubans, those with US dollars and those without. I was not aware of this. I can see down the track this could create enormous problems". Tom was also alerted to the potentially problematic situation:

> ... there is also in Cuba the creation of a two tier society, one which earns dollars from doing menial jobs in the tourist trade and the other which earns much less from doing all the rest of the important jobs that keep the country going. I'm really worried that none of the kids are going to want to be teachers, doctors, engineers. They're going to want to be waiters or leave Cuba.

Henrietta, an elderly American woman on the Cuba at the Crossroads tour, said:

> I have a much more concrete sense now of how, and how much, the embargo hurts
> Cubans, the mechanics by which it does so, and what the Cuban government and
> people have done in response. I will be writing to my political reps about ending
> the embargo and I do hope to work something of the Cuban experience into my
> healthcare talks. Meanwhile I've been mainly talking to friends and even clients
> about the insanity and destructiveness of the embargo.

In discussing what she has mostly learnt about and felt impacted by, Henrietta,
an elderly American woman on the Cuba at the Crossroads tour says:

> I have a much more concrete sense now of how, and how much, the embargo hurts
> Cubans, the mechanics by which it does so, and what the Cuban government and
> people have done in response. I will be writing to my political reps about ending
> the embargo and I do hope to work something of the Cuban experience into my
> healthcare talks. Meanwhile I've been mainly talking to friends and even clients
> about the insanity and destructiveness of the embargo.

Her account of how her tour had affected her is a striking example of the power of
this form of tourism to encourage participation in new social movements (Knoke
1988).

Furthermore, having learnt how the Cuban government is utilising tourism as
a developmental tool as part of its response to the embargo, Henrietta went on to
discuss some important issues relating to the embargo and tourism, demonstrating
her considered response to information gleaned through the seminars and project
visits:

> My main observations concern how the embargo is directly and indirectly
> dragging down the Cuban economy and how the answer they're seeking in
> tourism may be putting the entire culture at risk. I also observed what appears
> to be an interesting and potentially dangerous generational divide. Many of the
> people who were there for the revolution still seem to have great enthusiasm for
> it and great appreciation for what it accomplished. Younger people however who
> have no memory of pre-Castro days long for a better material life; not unlike
> the pre and post Depression generations in the US I guess. Tourism and the dual
> economy are introducing into Cuban life new economic and social disparities
> that could undermine the real, functional egalitarianism that has I believe made
> the revolution's successes possible.

As I described in Part 2, tourism was specifically introduced by the Cuban
government in response to the collapse of communism in Europe and subsequent
economic pressure from the US. It became the main money earner in the country
and was placed at the forefront of Cuba's development policy. After various

changes to legal structures, ordinary Cubans were allowed to hold foreign currency, to open US dollar bank accounts, and to spend US currency in the dollar stores. As part of these changes, it became legal for Cuban families to rent out rooms to tourists, upon the purchase of a license and a monthly fee. The license fee was increased significantly to dissuade local people from participating in this form of private enterprise and centralise the tourism market. More recently the US dollar was de-legalised.

This raises many questions as to how the Cuban government and people managed increasing social impacts of tourism. Henrietta highlighted one of the most explicit effects of tourism when she mentioned a "generational divide". As increasing numbers of foreign tourists entered Cuba, young Cubans took on a form of acculturation through expressing a desire to acquire the material wealth of foreigners. This was evident in their clothing, their accessories, and the music they listened to. The Cuban diaspora in Miami sent remittances and goods, such as music CDs that were then circulated on the black market. For most Cubans, overseas travel was not a possibility unless they received government funding for performing in music, arts or sports or if they received an invitation from an overseas citizen accompanied by the necessary funding. Otherwise, most Cubans' impressions about the West (especially those of Cuban youth) are formed from tourists and the media, and take it to be rich and glamorous. This illustrates the uneven benefits produced by globalisation that precipitate various kinds of frictions in local contexts (Appadurai 2001).

This generational divide manifested itself through tourism. Many Cuban youths were more inspired by the material wealth of capitalism than the values of the socialist revolution. International tourism being still relatively recent in Cuba, the Cuban government's responses to the social impacts of tourism continually changed as new social issues arose. The government decided the positive and negative effects of tourism on the economy and society in order to ensure the most successful outcome, while preserving the socialist value system fostered throughout the revolution. This dichotomy represented one of the challenges facing Cuba's attempts to reinsert itself into the globalising world and undergo sociopolitical change. If the next generation is not committed to the socialist achievements of the last four decades and capitalism eventually succeeds as the new orthodoxy, development-oriented tours might become obsolete. The social impacts of tourism highlight the risks involved and the pitfalls facing increased tourism as development. Tourism is an entity with a life of its own and thus it is very difficult to harness. This, in turn, means it is only possible to make provisional conclusions about Cuba and NGO tours in this book. What is clear is that the tours offered people a set of experiences that often led to transformative learning and critical perspectives of development.

The nature of the tourists' experiences was complex. There existed considerable contradictions between participants' backstage desires and their tourist inclinations. The tourists I came to know were motivated by moral imperatives to do something good and to learn from the grassroots about development issues. Indeed there were aspects of the tours that some participants were unhappy about precisely because they interfered with their education agenda, such as the cancellation of meetings, party line propaganda, and certain tourist attractions. There remained a tension between tourist expectations and the reluctance to identify oneself as a tourist. For example, while NGO tourists positioned themselves oppositionally to mainstream tourists because they engaged in a form of educational travel focusing on development issues, they still used tourist infrastructure and evidently held expectations similar to any other tourist about the level of service acceptable in a tourist facility. Indeed, Palmer (2005: 9) claimed that "a person may draw upon more than one identity depending upon their personal circumstances. These interweaving identities are like hats that can be changed to suit both the occasion and the mood of the person wearing them ... Therefore, identity is not a neutral concept". Travel is a perfect arena in which people can explore their own identity. Some tourism scholars have looked into:

> ... the connection between travel and evolving personal identities, arguing that the anticipation of, and narratives about, journeys on their return are tied into imagined 'performances' of the self (Desforges 2000). These 'performances' enable travellers to think of themselves as particular (or different) kinds of people (White and White 2004: 214); [morally superior to other tourists].

A moral imperative drives the increasing connection between globalised social movements, development and tourist concern with social change. The information exchanges taking place on tours helped Cuba to find a voice and in this sense retrieve some sort of agency within a global network. Such exchanges allowed Cuban people to become more of a presence within a wider geopolitical, economic and social world in which they had been subject to a decades long embargo excluding Cuba from world markets. The engagement of tourists and locals was an integral part of the development process as they shared their knowledge and expertise with each other. Subjective micro transformations of this kind led to potential multiplier effects beyond the tours themselves in the form of new networks being built, role models being created and positive social change being promoted.

Such experiences, combined with the goals of the NGOs and the intentions of the Cuban government, indicated several positive outcomes. First, outside of material support for the Cuban organisations meeting with the study tour groups and discussing development issues specific to Cuba, the positive outcomes consisted of imparting knowledge about Cuban realities, thereby facilitating solidarity between countries and potentially empowering Cuban people. This could impact on future international relations for Cuba in potentially assisting in future development (for example through tourists returning home to lobby their politicians about trade relations), which has been threatened by the US trade embargo.

Second, for the NGOs who organised study tours to Cuba, the positive outcomes consisted of participants learning about development efforts, disseminating this knowledge on their return home and perhaps becoming more committed to supporting the NGOs development efforts. Third, for the participants, the positive outcomes were partaking in a study tour that contributed money and possibly a sense of well-being to the local communities; gaining 'backstage' access not otherwise available to tourists; being educated while travelling; meeting like-minded people, identity reinforcement, feeling a sense of moral superiority, having an 'authentic' encounter, getting to know the locals. But, most importantly, they became actively engaged in activities linked to social solidarity and change. While NGO tourism is not a new social movement *per se*, it lends itself to particular outcomes associated with the power of new social movements. The level of impact on the tourists themselves through transformation led to agency that in turn enhanced social capital in Cuba.

Conclusion
Other Transformations: Rights-based
Development to Rights-based Tourism

Within emerging trends of tourism and development a culture of concern and the notion of 'moral responsibility' bring together rights-based development and 'new tourism'. Based on their conjunction we can conceive of it as a rights-based tourism. NGO study tours are based on a moral imperative to be contributing something 'good' to the communities visited. This imperative encompasses ethical considerations in so far as ethical stresses idealistic standards of right and wrong, such as codes of conduct within tourism. I use the term 'moral' to envisage the underpinnings of the convergence between development and tourism because it includes notions of ethical, right, virtuous, good. Such ideas of improvement (or at least careful avoidance of harm) correlate with rights-based development and the juxtaposition of these two elements leads to the creation of global networks, which is central to new social movements. Rights-based development, and its underlying moral imperative, aims to ensure that people have moral and legal entitlements that pertain specifically to basic well-being and dignity (Ljumgman 2005). In the context of tourism the notion of the moral has emerged from changing attitudes of the Western middle classes to embrace discourses of sustainability. In this book, tour participants take a stance on tourism issues that rests on minimising harm to the environment and cultures, indicating that in contrast, they generally perceive mass tourism as hedonistic and potentially damaging. NGO study tours are considered to be morally superior because they engage with local culture and incorporate a learning component. Participants indicate that they perceive themselves to be culturally sensitive or responsible while travelling. Much of their time in Cuba is spent in meetings with local people geared towards understanding the nuances of development and tourism in Cuba. For example, how the dilemma for the Cuban government between the economic benefits of ending the US embargo is weighed against the potential negative impacts that the influx of American tourists could have on Cuban culture. Participants also cite a strong concern for the disparity in wealth that is being created by tourism and discussed the tension between the benefits and the drawbacks of tourism as a development tool. In these ways they consider themselves to be engaging a more judicious form of tourism.

This book argues against Butcher's (2003) critique of "moral proscription and critical self awareness" in tourism, in which he argues the attempt to do something good in fact does the opposite as it creates new barriers between people and has a negative impact in the field of development. Butcher positions new moral tourisms as "a vessel into which we are encouraged to pour environmental angst and fears of globalisation. New Moral Tourists travel with a sense of personal mission, as tourism is recast as philanthropy towards the hosts and a 'unique experience' for the tourist" (2003: 139). It is far too broad to argue that *all* tourism takes on this moral dimension, or for that matter, that this increasing moralisation is simply jargon. But if there are increasing niches that bring about 'better' tourist practices that are not just rhetorical, as I have revealed, then further research is clearly warranted.

Global Exchange and Oxfam tours provide examples of the values and characteristics associated with the new tourism paradigm. The experiences analysed in this book can be seen to fit into this paradigm precisely because the positive nature of such experiences are perceived as constructive and morally superior. NGO study tours offer a convenient and secure 'off the beaten track' experience. These tourists also chose this style of holiday in order to distinguish themselves from other types of tourists, mainly mass tourists. Development-oriented tours are seen to help people to enact their moral choice as responsible alternative tourists. While illustrating how these tours exemplify the values and characteristics of new tourism, I also argue that the issue of morality is complex. In the context of NGO study tours in Cuba, notions of the moral are entwined with a rights-based tourism, which is also premised on a sense of responsibility towards people in need. Where Butcher challenges the positive assumptions of new tourism niches, I reveal its complexity as a culture of concern about damage to environment and culture that promulgates connections between people. This process is akin to new social movements and has positive impacts in the field of development. We can thus comprehend the convergence of development and tourism taking place in Cuba through a 'new tourism' lens; a 'moralisation of tourism' lens; and a 'rights-based' lens.

The focus of this book was on a tourism-development nexus with a moral underpinning, which involves Western tourists actively using their holidays to learn about development issues. It is through this process that tourists achieve a sense of agency. This clearly does not appeal to all tourists and I am not suggesting it will. But as this book indicated, there is an increasing tourism niche developing, which focuses on an exchange of knowledge and which undoubtedly has positive implications for a host country, leading as it does to the creation of global networks and solidarity.

Specific tourism niches can potentially lead to mutually beneficial relationships between development and tourism. NGO study tours suit tourists who seek to use their leisure time for personal growth and educational purposes. They afford tourists the unique opportunity to go beyond superficial interactions to gain an engaging interaction with the host culture through a program of meetings and project visits pertaining to development issues. Many participants claimed to gain a more intricate view of development-related issues through their study tour. In terms of cultural contact, most respondents indicated that they had far more personal contact with local Cuban people than if they had travelled on their own or with a mainstream tour operator. Many participants indicated that the contact with local people was not of a superficial nature and that they felt a deep connection because of the deep and enriching exchange of information and knowledge.

Development-oriented tours aim to move beyond typical touristic presentations of Cuba and counter the anti-Cuba rhetoric stemming from the US government and the Cuban diasporas. Cuba is all too often represented by discourses that shift between diverse idealistic visionary representations of a bastion of socialism fighting against consumerism and capitalism to dystopic representations of a socialist country frozen in the 1950s and a victim of the Cold War. The NGO study tours do more than feed into discourses of representation by providing Western tourists with opportunities to experience Cuba through a series of meetings that allow people to gain nuanced understandings of Cuban realities, both positive and negative.

Solidarity through tourism, in whatever form this emerges, can be considered an important tool for development agencies, social movements, and NGOs in terms of new and explicit ways of promulgating issues of rights, social justice and good governance. In this way, solidarity connects directly with rights-based development. Solidarity becomes important in the tourism context, because it is implicitly expressed as an objective of NGOs and explicitly expressed by tourists as a key motivation for participating on NGO study tours. Likewise it is expressly a political and developmental goal of the Cuban government. We can effectively envisage solidarity as a means for tourists to participate and act as agents of change in the development process. Indeed, it is a novel means through which Cuba has developed a way to partially transcend the economic and social constraints of the blockade. It acts as a new form of global coalition and interconnectedness that builds on previous alliances that have since dissolved, as was the case with the Soviet Union where Cuba engaged in cultural exchanges with nations who were politically sympathetic.

My ethnographic insights, suggest that participants were very keen to engage with Cuban people and learn about their development initiatives. Some people have contacted me since their tour because they wanted to share their post-tour efforts to maintain interest and keep abreast of Cuban issues. This indicates a keen move towards Cuban solidarity achieved through such activities as writing articles about Cuban social development initiatives; speaking to friends, colleagues or giving formal addresses to large groups; becoming more active in campaigning local politicians or joining the Cuba Friendship Society. These post-tour actions illuminate individual expressions of "desire", in Kapoor's (2005) sense, at least in so far as the tourists aim to be part of social movements and development activities. It is in these ways that people also actively overcome entrenched Western "complicity" over Third World underdevelopment and poverty. While they may wish to be seen as benevolent tourists, they can at the very least be seen to be more active in their solidarity with Cuba post-tour, in direct contrast to most Western tourism to the Third World.

It can be argued that this is a form of tourism that attempts to overcome Western complicity in Third World poverty. We can take this argument one step further and say that participants in development study tours are, through their solidarity efforts, agents of a rights-based form of development. The tours have the power to affect participants in positive ways that encourage people to be more active in a growing social movement of foreigners supporting Cuba. Tourists become agents of development by supporting the development projects of the NGO they travelled with and supporting Cuban solidarity. Thus, both the Cuban organisations and NGOs are meeting their objectives in conducting these tours. What is produced here is rights-based tourism.

Rights-based tourism, as exemplified by the groups going to Cuba, requires new moral tourism in order to happen because of its implications for development. While I have used Butcher's terms I differ from him in asserting that moral imperatives driving new forms of tourism necessitate research into their potential for development. This tourism-development nexus is not just about material and financial exchanges but also about intellectual and affective elements that are exchanged and developed. This fits within a rights-based development framework where the notion of well-being is achieved by broadly conceived notions of political and moral support that is not just about money. Tours contribute (on different scales) to the development of dignity and well-being because NGO tourists leave Cuba with a sense of connection and the commitment to support development in Cuba.

Development-oriented tours with NGOs are of an active participatory and educational nature. It is these distinctive qualities that explain why participation in an NGO study tour can increase network ties. By interaction with fellow tour participants, local officials and local people the establishment of network ties is facilitated. Although the tours are only of several weeks duration, despite this short immersion the experience is intense enough to have an impact on the tourists. The tours allowed participants many opportunities to exchange information about networks and to develop ties that would never have occurred had they not participated in a development tour. Tours of this kind would be expected to draw together like-minded people, and it thus makes sense that tours enable the exchange of ideas and establishment of network ties between the tourists themselves. In this sense the notion of transformation is qualified by notions of affirmation. These new network ties have the potential to encourage participation in social movements. Many tour participants acknowledged that these tours promote further support or activism. For example participants stated they will lobby their politicians, give presentations about their experiences and write articles. Hence development study tours, by virtue of their intense social experience, lead to the creation of networks and rights-based tourism.

The sharing of ideas with local people, fellow tour members and with people back home can promote greater understanding of issues regarding systems of aid and thereby increase the contributions they make to development. It is this impact that leads to the 'transformation' from tourist to agent of development because tour participants are engaged in more long-term intellectual and instrumental ways that transcend the tour itself. I have discussed that notions of touristic transformation are complex forms of identification from rejection of mass tourism to refutation of neoliberal globalisation. The tourists demonstrated identification with certain interests (for example ecology and organic farming). It is often the 'peasant' that is valued within these tourist encounters, not the musician or the entertainer as is usually the case within tourism experiences. Transformation is qualified by the tourists' affirmation of values espoused in the West – such as social equity, collective community – but no longer considered valid with the collapse of the Soviet Empire. Cuba forms an example of a model that tourists are compelled to study, support, and promote.

Clearly there needs to be further research about the contributions of this form of tourism to a rights-based development. To what extent and how consistently such tours improve the social and cultural environment of local people by enhancing their sense of well-being and dignity remains unclear. But the argument that international networks and solidarity are created leads us to ask such questions. This book has argued for the recognition of the potential power of development tours to provide the means to establish relationships that extend beyond the brief tour experience itself. This book, more than anything else, demonstrates that if a development tour is found to encourage or intensify solidarity with Cuba or social movement participation in some form then the results could be used to promote rights-based tourism as a means of encouraging organised social action and rights-based development.

Bibliography

Abercrombie, N., S. Hill, B. Turner (eds) (1994) *The Penguin Dictionary of Sociology*. London: Penguin Books.

Adams, K. (1991) 'Distant Encounters: Travel Literature and Shifting Images of the Toraja of Sulawesi, Indonesia', *Terrae Incognitae* 16: 84-92.

Adams, W. (1990) *Green Development: Environment and Sustainability in the Third World*. London: Routledge.

Adler, J. (1989) 'Origins of Sightseeing', *Annals of Tourism Research* 16, 1: 7-9.

Ahmed, Z., F. Krohn, V. Heller (1994) 'International Tourism Ethics as a Way to World Understanding', *Journal of Tourism Studies* 5, 2: 36-44.

Allen, J. and D. Massey (eds) (1995) *Geographical Worlds*. Milton Keynes: Open University Press.

Amir, Y. and R. Ben-Ari (1985) 'International Tourism, Ethnic Contact and Attitude Change', *Journal of Social Issues* 41, 3: 105-115.

Appadurai, A. (1990) 'Disjuncture and Difference in the Global Political Economy'. In *Global Culture*. M. Featherstone (ed.) London: Sage.

Appadurai, A. (1996) *Modernity at large: Cultural dimensions of globalisation*. Minneapolis: University of Minnesota Press.

Appadurai, A. (2001) *Globalization*. London: Duke University Press.

Ashley, C., C. Boyd. and H. Goodwin (2000) 'Pro-Poor Tourism: Putting Poverty at the Heart of the Tourism Agenda', *Natural Resource Perspectives* 51, March 2000.

Ashley, C., D. Roe and H. Goodwin (2001) *Pro-Poor Tourism Strategies: Making Tourism Work for the Poor*. Nottingham: The Russell Press.

Baker, C. (1997) 'Membership Categorization and Interview Accounts'. In *Qualitative Research: Theory, Method and Practice*. D. Silverman (ed.) London: Sage.

Baran, P. (1957) *The Political Economy of Growth*. New York: Monthly Review Press.

Barkan, S., S. Cohn and W. Whitaker (1995) 'Beyond Recruitment: Predictors of Differential Participation in a National Antihunger Organisation', *Sociological Forum* 10: 113-132.

Baudrillard, J. (1988) *America*. London: Verso.

Bauman, Z. (1993) *Post-modern Ethics*. Oxford: Blackwell.

Bauman, Z. (1997) *Postmodernity and its discontents*. Cambridge: Polity.

Benjamin, M. (1997) *Cuba: Talking About Revolution*. Melbourne: Ocean Press.

Benjamin, M. and A. Freedman (1992) *Bridging the Global Gap: A Handbook to Linking Citizens of the First and Third Worlds.* Washington DC: Seven Locks Press.

Bennett, O., D. Roe and A. Caroline (1999) *Sustainable Tourism and Poverty Elimination Study,* a report to the Department for International Development: London.

Berlyne, D. (1962) 'New Directions in Motivation Theory', *Anthropology and Human Behaviour,* T. Gladwin and W.C. Sturtevant (eds) Washington DC: Anthropological Society of Washington.

Bleasdale, S. and Tapsell, S. (1994) 'Contemporary Efforts to Expand the Tourist Industry in Cuba: The Perspective from Britain'. In *Tourism: The State of the Art.* A.V. Seaton. Brisbane: John Wiley and Sons.

Boorstin, D. (1964) *The Image: A Guide to Pseudo-Events in America.* New York: Harper.

Boorstin, D. (1969) *The Image: A Guide to Pseudo-Events in America.* New York: Vintage Books.

Bourdieu, P. (1984) *Distinction: A Social Critique of the Judgement of Taste.* Trans. Richard Nice. Cambridge: Harvard University Press.

Boyte, H. (1980) *The Backyard Revolution: Understanding the New Citizen Movement.* Philadelphia: Temple University Press.

Bramwell, B., I. Henry, G. Jackson and J. van der Straaten (1996) 'A Framework for Understanding Sustainable Tourism Management'. In *Sustainable Tourism Management: Principles and Practice.* B. Bramwell, I. Henry, G. Jackson, A Prat, G. Richards and J. van der Straaten (eds) Tilberg: Tilberg University Press.

Britton, S. (1982) 'The Political Economy of Tourism in the Third World', *Annals of Tourism Research* 9: 331-58.

Britton, S. (1989) 'Tourism. Dependency and development: A mode of analysis'. In *Ambiguous Alternative: Tourism in Small Developing Countries.* S.G. Britton and W.C. Clarke (eds) Suva: University of the South Pacific.

Broad, S. (2002) *Anything but a Holiday? Volunteer Tourism and the Gibbon Rehabilitation Project, Thailand,* Unpublished Doctoral Thesis, University of Newcastle.

Brown, D. (1998) 'In Search of an Appropriate form of Tourism for Africa: Lessons from the past and suggestions for the future', *Tourism Management* 19: 237-245.

Bruner, E. (1991) 'Transformations of Self in Tourism', *Annals of Tourism Research* 18: 238-250.

Bruner, E. (2005) *Culture on Tour.* Chicago: University of Chicago Press.

Butcher, J. (2003) *The Moralisation of Tourism: Sun, Sand ... and Saving the World?* London: Routledge.

Butler, R. (1999) 'Sustainable Tourism: A State of the Art Review', *Tourism Geographies* 1, 1:7-25.

Campbell, C. (2001) *The Changing Face of Inner City Havana.* Ottawa: IDRC.

Chambers, R. (2005) 'Critical Reflections of a development nomad'. In *A Radical History of Development Studies: Individuals, Institutions and Ideologies*. U. Kothari (ed.) London: Zed Books.

Clifford, J. (1988) *The Predicament of Culture: Twentieth Century Ethnography, Literature and Art*. Cambridge: Harvard University Press.

Clifford, J. (1997) *Routes: Travel and Translation in the Late Twentieth Century*. Cambridge: Harvard University Press.

Cohen, E. (1972) 'Toward a Sociology of International Tourism', *Social Research* 39, 1: 164-182.

Cohen, E. (1974) 'Who is a Tourist? A Conceptual Review', *Sociological Review* 22: 27-53.

Cohen, E. (1979a) 'Rethinking the Sociology of Tourism', *Annals of Tourism Research* 6, 1: 18-35.

Cohen, E. (1979b) 'A Phenomenology of Tourist Experiences', *Sociology* 13: 179-202.

Cohen, E. (1987) '"Alternative Tourism" – A Critique', *Tourism Recreation Research* 6, 1: 18-35.

Cohen, E. (1988a) 'Traditions in the Qualitative Sociology of Tourism', *Annals of Tourism Research* 15: 29-46.

Cohen, E. (1988b) 'Authenticity and Commoditisation in Tourism', *Annals of Tourism Research* 15: 371-386.

Cohen, E. (1989) 'Primitive and Remote: Hill Tribe Trekking in Thailand', *Annals of Tourism Research* 16, 1: 30-59.

Cohen, E. (1995) 'Contemporary Tourism – trends and challenges: Sustainable authenticity or contrived post-modernity?'. In *Change in Tourism: People, Places, Processes*. R. Butler and D. Pearce (eds) London: Routledge.

Cornwall, A. and C. Nyamu-Musembi (2004) 'Putting the "rights-based approach" to development into perspective', *Third World Quarterly* 25, 8. 1415-1437.

Crewe, E. and E. Harrison (2002) *Whose Development? An Ethnography of Aid*. London: Zed Books.

Crick, M. (1988) 'Sun, Sex, Sights, Savings and Servility', *Criticism, Heresy and Interpretation* 1: 37-76.

Crick, M. (1989) 'Representations of International Tourism in the Social Sciences: Sun, Sex, Sights, Savings and Servility', *Annual Review of Anthropology* 18: 307-344.

Crompton, J. (1979) 'Motivations for Pleasure Vacation', *Annals of Tourism Research* Oct/Dec.

Crompton, J. and P. Ankomah (1993) 'Choice set Propositions in Destination Decisions', *Annals of Tourism Research* 20, 3: 461-476.

Crompton, R. (1993) *Class and Stratification*. Oxford: Polity Press.

Cross, P. (1992) 'Soviet Perestroika: The Cuban Effect', *Third World Quarterly* 13, 1: 143-158.

Crush, J. (1995) *The Power of Development*. London: Routledge.

D'Amore, L. (1988) 'Tourism – The World's Peace Industry', *Journal of Travel Research* 27, 1: 35-40.

Dag Hammarskjold Foundation (1975) *What Now? Another Development*. A Special Issue of Development Dialogue. Uppsala: The Foundation.

Dann, G. (1977) 'Anomie, Ego-enhancement and Tourism', *Annals of Tourism Research* 15: 269-283.

Dann, G. (1981) 'Tourist Motivation: An Appraisal', *Annals of Tourism Research* 8, 4: 187-219.

Dann, G. (1999) 'Writing Out the Tourist in Space and Time', *Annals of Tourism Research* 26: 159-187.

Dann, G. (2000) *Push-Pull Factors. Encyclopedia of Tourism*. J. Jafari (ed.) London: Routledge.

Day-Lewis, C. (1983) *Virgil: The Eclogues, The Geogrics*. London: Oxford University Press.

De Rivero, O. (2001) *The Myth of Development*. London: Zed Books.

Deleuze, G. and F. Guattari (1988) *A Thousand Plateaus. Capitalism and Schizophrenia*. Brian Massumi, Translation. London: Athlone Press.

Denzin, Norman K. (1970) *The research act: A theoretical introduction to sociological methods*. Chicago: Aldine Pub. Co.

Department for International Development (1999) *Tourism and Poverty Elimination: Untapped Potential*. London: DFID.

Derrida, J. (1974) *Of Grammatology*. Baltimore: Johns Hopkins University Press.

Desforges, L. (1998) 'Checking Out the Planet: Global Representations/Local Identities and Youth Travel'. In *Cool Places: Geographies of Youth Culture*. T. Skelton and G. Valentine (eds) London: Routledge.

Desforges, L. (2000) 'Travelling the World: Identity and Travel Biography', *Annals of Tourism Research* 27, 4:926-945.

Dollar, D. and A. Kraay (2000) *Growth is Good for the Poor*. Washington DC: World Bank.

Donnelly, J. (1989) *Universal Human Rights in Theory and Practice*. New York: Cornell University Press.

Eco, U. (1986) 'Travels in Hyperreality'. In *Travels in Hyperreality: Essays*. San Diego: Harcourt Brace Jovanovich.

Edelman, M. and A. Haugerud (2004) *The Anthropology of Development and Globalisation. From Classical Political Economy to Contemporary Neoliberalism*. Melbourne: Blackwell Publishing.

Escobar, A. (1995) *Encountering Development: The Making and Unmaking of the Third World*. Princeton: Princeton University Press.

Escobar, A. (1997) 'Anthropology and Development', *International Social Science Journal* 154: 497-515.

Escobar, A. (2005) 'Imagining a Post-Development Era'. In *The Anthropology of Development and Globalisation: From Classical Political Economy to Contemporary Neoliberalism*. M. Edelman and A. Haugerud (eds) Oxford: UK.

Esteva, G. (1988) 'El desastre agrícola: Adiós al México imaginario'. *Comercio Extrerio* 38, 8:662-72.

Esteva, G. (2001) *Development. In the Development Dictionary: A Guide to Knowledge as Power*. Wolfgang Sachs (ed.) London: Zed Books.

Farrell, B. and L. Twining-Ward (2004) 'Reconceptualizing Tourism', *Annals of Tourism Research* 31, 2: 274-295.

Ferguson, J. (1990) *The anti-politics machine "development", depoliticization and bureaucratic power in Lesotho*. New York: Cambridge University Press.

Ferguson, J. (1999) *Expectations of modernity: Myths and meanings of urban life on the Zambian Copperbelt*. Berkeley: University of California Press.

Ferguson, N. (2002) *Empire: The Rise and Demise of the British World Order and the Lessons for Global Power*. New York: Basic Books.

Figueras, M. (2002) pers. comm. Advisor to the Minister of Tourism, Cuba.

Fisher, W. (1997) 'DOING GOOD? The Politics and Anitpolitics of NGO Practices', *Annual Review of Anthropology* 26: 439-64.

Foreign Investment Act pamphlet (1995) Consultores Asociados SA: Cuba.

Foucault, M. (1972) *The Archaeology of Knowledge*. New York: Harper Colophon Books.

Foucault, M. (1980) *Power/Knowledge*. London: Tavistock.

Frank, A. (1967) *Capitalism and Underdevelopment in Latin America*. London: Monthly Review Press.

Gardener, K. and D. Lewis (1996) *Anthropology, Development and the Postmodern Challenge*. London: Pluto Press.

Gasperini, L. (1999) *The Cuban Education System: Lessons and Dilemmas*. World Bank Report.

Giddens, A. (1989) *Sociology*. Cambridge: Polity Press.

Giddens, A. (1991) *Modernity and Self Identity: Self and Society in the Late Modern Age*. Cambridge: Polity.

Giddens, A. (2002) 'Living in a Post-Traditional Society'. In *Reflexive Modernisation: Politics, Tradition and Aesthetics in the Modern Social Order*. U. Beck, A. Giddens and S. Lash (eds) Cambridge: Polity Press.

Goffman, E. (1959) *The Presentation of Self in Everyday Life*. New York: Doubleday.

Gonsalves, P. (1983) 'Divergent Views: Convergent paths: Towards a Third World critique of tourism', *Contours* 6, 3/4: 8-14.

Goodfellow, D. (1991) 'Hosts versus guests – the desires and concerns of the passengers on Society Expedition's ship "World Discoverer"'. In *Ecotourism incorporating the Global Classroom: International Conference Papers*. B. Weiler (ed.) Australia: Bureau of Tourism Research.

Gordon, J. (1997) *Cuba's Entrepreneurial Socialism*. The Atlantic Online, http://www.theatlantic.com/issues/97jan/Cuba.

Gottlieb, A. (1982) 'American's Vacations', *Annals of Tourism Research* 9: 165-187.

Graburn, N.H.H. (1980) 'Teaching the Anthropology of Tourism', *International Social Science Journal* 32: 56-68.

Graburn, N.H.H. (1983) 'The Anthropology of Tourism', *Annals of Tourism Research* 10: 9-33.

Graburn, N.H.H. (1989) 'Tourism: The Sacred Journey'. In *Hosts and Guests: The Anthropology of Tourism*, 2nd edition. V. Smith (ed.) Philadelphia: University of Pennsylvania Press.

Gray, H. (1970) *International Travel – International Trade*. Lexington: Heath Lexington.

Guha, R. (1983) 'The Prose of Counter-Insurgency'. In R. Guha, *Subaltern Studies* 11, New Delhi: Oxford University Press.

Guha, R. (1988) 'Preface'. In *Selected Subaltern Studies*. R. Guha and G. Spivak. New York: Oxford University Press.

Guha, R. (1996) 'The Small Voice of History'. In *Subaltern Studies* 9. R. Guha. New Delhi: Oxford University Press.

Habermas, J. (1981) 'New Social Movements', *Telos* 49: 33-37.

Hall, C. and B. Weiler. (1992) 'Introduction. What's Special about Special Interest Tourism'. In *Special Interest Tourism*. New York: Wiley.

Hall, D.R. (1984) 'Foreign tourism under socialism: The Albanian "Stalinist" model', *Annals of Tourism Research* 11: 539-555.

Hall, D.R. (1992) 'Tourism Development in Cuba'. In *Tourism in the Less Developed Countries*. D. Harrison, (ed.) London: Halsted Press.

Hall, S. (1992a) 'The Question of Cultural Identity'. In *Modernity and its Future*. S. Hall, D. Held and T. McGrew (eds) Oxford: Polity Press.

Hall, S. (1992b) 'The West and the Rest: discourse and power'. In *Formation of Modernity*. S. Hall and B. Gieben (eds) Polity Press: Oxford.

Hall, S. (2002) 'The West and the Rest: Discourse and Power'. In *Development: A Cultural Studies Reader.* S. Schech and J. Haggis (eds) Oxford: Blackwell Publishers.

Hamm, B. (2001) 'A Human Rights Approach to Development', *Human Rights Quarterly* 23: 1005-1031.

Hannerz, U. (1996) *Transnational Connections: culture, people, places*. London: Routledge.

Hardy, D. (1990) 'Socio-cultural Dimensions of Tourism History', *Annals of Tourism Research* 17, 4: 541-555.

Hardt, M. and A. Negri. (2000) *Empire*. London: Harvard University Press.

Harvey, D. (1989) *The Conditions of Postmodernity*. Oxford: Blackwell Publishers.

Hinch, T. (1990) 'Cuban Tourism: Its Re-emergence and Future', *Tourism Management* 11, 3: 214-26.

Hinch, T. and R. Butler (1996) 'Indigenous Tourism: A Common Ground for Discussion'. In *Tourism and Indigenous Peoples*. R. Butler and T. Hinch (eds) London: International Thomson Business Press.

Hirschman, A. (1958) *Strategy of Economic Development.* New Haven: Yale University Press.

Hobsbawm, E. and T. Ranger (eds) (1983) *The Invention of Tradition.* New York: Cambridge University Press.

Hodge, P. (2004) *Development Issues. Volunteer Work Overseas: For Australians & New Zealanders.* Newcastle: Global Exchange.

Hollander, P. (1981) *Political Pilgrims: Travels of Western Intellectuals to the Soviet Union, China and Cuba 1928-1978.* Oxford: Oxford University Press.

Hollander, P. (1986) 'Political Tourism in Cuba and Nicaragua', *Society* May/June: 28-37.

Holmes, D. (1998) 'Tourist Worlds as Monoculture: Learning from the Gold Coast'. In *Tourism, Leisure, Sport: Critical Perspectives.* D. Rowe and G. Lawrence (eds) Rydalmere: Hodder Education.

Hunter, C. (1995) 'On the Need to Re-Conceptualise Sustainable Tourism Development', *Journal of Sustainable Tourism* 3, 3: 155-165.

Hunter, C. (1997) 'Sustainable Tourism as an Adaptive Paradigm', *Annals of Tourism Research* 24: 850-867.

IDRC. *IDRC in Cuba* in IDRC Country Profiles, http://www.idrc.ca/reports. As posted in May 2003.

Illich, I. (2001a) 'Development as Planned Poverty'. In *The Post-Development Reader.* Majid Rahnema and Victoria Bawtree (eds) London: Zed Books.

Illich, I. (2001b) 'Needs'. In *The Development Dictionary: A Guide to Knowledge as Power.* W. Sachs (ed.) London: Zed Books.

Inglehart, R. (1977) *The silent revolution: Changing values and political styles among Western publics.* Princeton, NJ: Princeton University Press.

Jafari, J. (1987) 'Tourism Models: The Sociocultural Aspects', *Tourism Management* 8: 151-159.

Jameson. F. (2002) *A Singular Modernity: Essay on the Ontology of the Present.* London: Verso.

Jimenez, A. (1990) *Guide to Varadero.* Italy: Gianni Constantino.

Jolly, M. (1992) 'The Contemporary Pacific', *A Journal of Island Affairs* 4, 1: 49-72.

Kakwani, N. and E. Pernia (2000) 'What is Pro-poor Growth?', *Asian Development Bank* 18, 1.

Kalish, A. (2001) *Tourism as Fair Trade: NGO Perspectives.* London: Tourism Concern.

Kapoor, I. (2005) 'Participatory Development, Complicity and Desire', *Third World Quarterly* 26, 8: 1203-1220.

Kelly, I. (1997) 'Study Tours: A Model for "Benign" Tourism?', *Journal of Tourism Studies* 8, 1: 42-51.

Klandermans, B. and D. Oegema. (1987) 'Potentials, Networks, Motivations and Barriers: Steps Towards Participation in Social Movements', *American Sociological Review* 52: 519-531.

Klasen, S. (2001) *In Search of the Holy Grail: How to Achieve Pro-Poor Growth?* A Paper Commissioned by Deutsche Gesellschaft für Technische Zusammenarbeit (GTZ).

Klein, N. (2001) *No Logo.* London: Flamingo.

Klein, N. (2004) 'Reclaiming the Commons'. In *A Movement of Movements: Is another world really possible?* Tom Mertes (ed.). London: Verso.

Knauft, B. (2002) 'Critically Modern: An Introduction'. In *Critically Modern: Alternatives, Alterities, Anthropologists.* B. Knauft (ed.) Bloomington: Indiana University Press.

Knoke, D. (1988) 'Incentives in Collective Action Organisations', *American Sociological Review* 53: 311-329.

Kothari, U. and M. Minogue (2002) *Development Theory and Practice: Critical perspectives.* Hampshire: Palgrave.

Krippendorf, J. (1987) *The Holiday Makers: Understanding the Impact of Leisure and Travel.* Oxford: Butterworth Heinemann.

Lash, S. and J. Urry. (1987, 1994) *Economies of Signs and Space.* London: Sage.

Lattas, A. (1993) 'Essentialism, Memory and Resistance: Aboriginality and the Politics of Authenticity', *Oceania* 63: 240-267.

Laxson, J. (1991) 'How "we" see "them" tourism and Native Americans', *Annals of Tourism Research* 18, 3: 365-391.

Lea, J. (1988) *Tourism and Development in the Third World.* London: Routledge.

Lett, J. (1983) 'Lucid and Liminoid aspects of charter yacht tourism in the Caribbean', in *Annals of Tourism Research* 10: 35-56.

Lewis, A.W. (1955) *The Theory of Economic Growth.* London: Allen & Unwin.

Lichterman, P. (1996) *The Search for Political Community: American Activists Reinventing Commitment.* Cambridge: Cambridge University Press.

Linnekin, J. (1983) 'Defining Tradition: Variations on the Hawaiian Identity', *American Ethnologist* 10, 2: 241-252.

Lipman, G. (2004) *Tourism, Aviation and Poverty Reduction – Need for New Strategies.* Special Report to Secretary General WTO.

Ljumgman, C. (2005) 'A Rights Based Approach to Development'. In *Methods for Development Work and Research – A New Guide for Practitioners.* B. Mikkelsen (ed.) New Delhi: Sage.

Lobe, J. *Learn From Cuba, Says World Bank,* Inter Press Service. http://www.oneworld.org/ips2/apr01/00_21_003.html. As posted in August 2002.

MacCannell, D. (1973) 'Staged Authenticity: Arrangements of Social Space in Tourist Settings', *American Journal of Sociology* 79, 3: 589-603.

MacCannell, D. (1976) *The Tourist: A New Theory of the Leisure Class.* London: Macmillan.

MacCannell, D. (1992) *Empty Meeting Grounds: the Tourist Papers.* London: Routledge.

MacMichael, P. (1996) *Development and Social Change.* London: Sage.

Madruga, A. (2000) 'A Poor Country Does Not have to Leave its People Defenceless', *Granma Internacional* April 19.

Marcus, G. (1995) 'Ethnography in/of the World System: The Emergence of Multi-Sited Ethnography', *Annual Review of Anthropology* 24: 95-117.

Marks, S. (2003) *The Human Rights Framework for Development: Seven Approaches.* Working Paper No. 18. Francois-Xavier Bagnoud Centre for Health and Human Rights, Harvard University. As posted: http://www.cdra.org.za, November 2005.

Marx, K. (1965) *Pre-capitalist economic formations*, translated by Jack Cohen, E.J. Hobsbawm (ed.) New York: International Publishers.

Masberg, B. and L. Silverman (1996) 'Visitor Experiences at Heritage Sites: A Phenomenological Approach', *Journal of Travel Research* 34, 4: 20-25.

Matless, D. (1995) 'The Art of Right Living: Landscape and Citizenship, 1918-39'. In *Mapping the Subject: Geographies of Cultural Transformation*, S. Pile and N. Thrift (eds). London: Routledge.

McAdam, D. and D. Rucht (1993) 'The Cross-National Diffusion of Movement Ideas', *AAPSS Annals* 528: 56-74.

McCabe, S. and E. Stokoe (2004) 'Place and Identity in Tourists' Accounts', *Annals of Tourism Research* 31, 3: 601-622.

McGehee, N. (2002) 'Alternative Tourism and Social Movements', *Annals of Tourism Research* 29, 1: 124-143.

McKay, A. (1997) 'Poverty Reduction through Economic Growth: Some Issues', *Journal of International Development* 9, 4: 665-673.

McKinley, J.C. (1999) 'In Cuba's Crippled Economy, The Only Goal is Survival', *The New York Times* 11 January.

McMichael, P. (2004) *Development and Social Change: A Global Perspective*, 3rd edition. California: Sage.

Miller, D. (ed.) (1995) *Acknowledging Consumption: A Review of New Studies.* London: Routledge.

Morris, M. (1988) 'At Henry Parkes Motel', *Cultural Studies* 2, 1: 1-47.

Mowforth, M. and I. Munt (1998, 2003) *Tourism and Sustainability: New Tourism in the Third World.* London: Routledge.

Muetzelfeldt, M. (2002) 'Fieldwork at Home', in *Doing Fieldwork: Eight personal accounts of social research.* J. Perry (ed.) Sydney: UNSW Press in association with Deakin University Press.

Munt, I. (1994a) 'The 'Other' Postmodern Tourism: Culture, Travel and the New Middle Classes', *Theory, Culture and Society* 11: 101-123.

Munt, I. (1994b) 'Eco-tourism or Ego-tourism?' *Race and Class* 36, 1: 49-60.

Nash, D. (1979) 'An Aristocratic Tourist Culture: Nice', *Annals of Tourism Research* 6, 1: 61-75.

Nash, D. (1981) 'Tourism as an anthropological subject', *Current Anthropology* 22, 5: 461-481.

Nash, D. (1989) 'Tourism as a form of imperialism'. In *Hosts and Guests: The Anthropology of Tourism.* V. Smith (ed.) Philadelphia: University of Pennsylvania Press.

Nash, D. (1996) *Anthropology of Tourism.* Oxford: Pergamon.

Nederveen Pieterse, J. (2004) *Globalisation or Empire*. New York: Routledge.

Nightingale, D.J. and J. Cromby. (eds) (1999) *Social constructionist psychology: A critical analysis of theory and practice*. Buckingham: Open University Press.

Norton, R. (1993) 'Culture and Identity in the South Pacific: A Comparative Analysis', *Man* 28, 4: 741-759.

Noy, C. (2004) 'This Trip Really Changed Me: Backpackers' Narratives of Self-Change', *Annals of Tourism Research* 31, 1: 78-102.

O'Grady, R. (1982) *Tourism in the Third World*. New York: Orbis.

Oramas, J. (2001) 'Foreign Debt Remains at $11 billion USD', *Granma Internacional* April 12.

Orams, M. (1997) 'The Effectiveness of Environmental Education Can We Turn Tourists into Greenies?', *Progress in Tourism and Hospitality Research* 3: 295-306.

Overington, C. (2003) 'Living in Castro's Shadow', *The Age* May 17.

Palmer, C. (2005) 'An Ethnography of Englishness: Experiencing Identity through Tourism', *Annals of Tourism Research*, 32, 1: 7-27.

Parajuli, P. (2001) 'Power and Knowledge in Development Discourse: New Social Movements and the State of India', pp. 258-88. In *Democracy in India*. Gopal Jayal (ed.) New Delhi: Oxford University Press.

Pearce, D. (1987) *Tourism Today: A Geographical Analysis*. London: Longman.

Pearce, D. (1995) *Tourism Today: A Geographical Analysis*. Harlow: Longman Scientific and Technical.

Perry, J.M. et al. (1997) Cuban Tourism, Economic Growth and the Welfare of the Cuban Worker, Cuba in Transition: Volume 7: Papers and Proceedings of the Seventh Annual Meeting of the Association for the Study of the Cuban Economy (ASCE): 1997, August 7-9 Florida.

Pfaff, S. (1995) *Collective Identity, Informal Groups and Revolution Mobilisation*, Unpublished Master's Thesis, University of North Carolina.

Pizam, A., N. Uriely, A. Reichel (2000) 'The Intensity of Tourist-Host Social Relationships and its Effects on Satisfaction and Change of Attitudes: The Case of Working Tourists in Israel', *Tourism Management* 21, 4: 395-406.

Poon, A. (1989) 'Competitive Strategies for a new tourism'. In *Progress in Tourism, Recreation and Hospitality Management*. London: Belhaven.

Poon, A. (1993) *Tourism, Technology and Competitive Strategies*. Wallingford: CAB International.

Prosser, R. (1994) 'Societal Change and Growth in alternative tourism'. In *Ecotourism a Sustainable Option?* E. Cater and G. Lowman (eds) Chichester: John Wiley and Sons.

Rahnema, M. (1997) 'Introduction'. In *The Post-Development Reader*. M. Rahmena and V. Bawtree (eds) London: Zed Books.

Ravallion, M. (1997) 'Good and Bad Growth: the Human Development Reports', *World Development* 25, 5: 631-638.

Roe, D. and P. Urquhart (2001) 'Pro-Poor Tourism: Harnessing the World's Largest Industry for the World's Poor', *Opinion: World Summit on Sustainable Development* May 2001.

Roggenbuck, J. W., R.J. Lookis and J. Dadostino (1990) 'The Learning Benefits of Leisure', *Journal of Leisure Research* 22, 2: 112-124.

Rojek, C. (1993) 'De-differentiation and Leisure' *Society and Leisure* 16, 1: 15-29.

Ross, S. and G. Wall (1999b) 'Evaluating Ecotourism: The Case of North Sulawesi, Indonesia' *Tourism Management* 20, 6: 673-682.

Rostow, W.W. (1960) *The Stages of Economic Growth: A Non-Communist Manifesto*. Cambridge: Cambridge University Press.

Sachs, W. (1992) 'One World'. In *The Development Dictionary*. W. Sachs (ed.) London: Zed Books.

Sachs, W. (1999, 2001) *The Development Dictionary: A Guide to Knowledge as Power*. London: Zed Books.

Saney, I. (2004) *Cuba: A Revolution in Motion*. London: Zed Books.

Sarantakos, S. (1993) *Social Research*. South Melbourne: Macmillan Education.

Schuurman, F. (1993) *Beyond the Impasse: New directions in development theory*. London: Zed Books.

Sen, A. (1999) *Development as Freedom*. Oxford: Oxford University Press.

Shields, R. (1990) *Places on the Margin*. London: Routledge.

Shils, E. (1981) *Tradition*. London: Faber.

Short, J.R. (1991) *Imagined Country: Society, Culture and Environment*. London: Routledge.

Sinclair, M. and M. Thompson (2001) *Cuba, Going against the Grain: Agricultural Crisis and Transformation*. Boston: Oxfam America.

Slim, H. (2002) 'Making Moral Ground. Rights as the Struggle for Justice and the Abolition of Development'. In *PRAXIS, The Fletcher Journal of Development Studies* Vol XVII. As posted: http://fletcher.tufts.edu/praxis/xvii/Slim.pdf, December 2003.

Smilie, I. (1997) 'NGOs and development assistance: A change in mind-set?' *Third World Quarterly* 18, 3: 563-577.

Smith, M. and R. Duffy (2003) *The Ethics of Tourism Development*. London: Routledge.

Smith, V. (ed.) (1978) *Hosts and Guests: The Anthropology of Tourism*. Philadelphia: University of Pennsylvania.

Smith, V. (ed.) (1989) *Hosts and Guests: The Anthropology of Tourism*, 2nd edition. Philadelphia: University of Pennsylvania.

Sofield, T. (2003) *Empowerment for Sustainable Tourism Development*. Sydney: Pergamon.

Solman, P. (2001) *Capitalism in Cuba*. Online News Hour, http://www.pbs.org/newshour/bb/latin_america/july-dec01/cuba. As posted on August 2002.

Spencer, R. (1999) *'Giving Something Back'? An Analysis of Myth, Representation and Narrative in Tourism Discourse*, Unpublished Honours Thesis, University of Newcastle.

Spivak, G. (1988) 'Can the Subaltern Speak?'. In C. Nelson and L. Grossberg, *Marxism and the Interpretation of Culture*. Urbana: University of Illinois Press.

Spivak, G. (1996) 'Subaltern Talk: Interview with the Editors 1993-4'. In *The Spivak Reader*. D. Landry and G. Maclean (eds) New York: Routledge.

Stabler, M. and B. Goodall. (1996) 'Environmental Auditing in Planning for Sustainable Island Tourism'. In *Sustainable Tourism in Islands and Small States: Issues and Policies*. L. Briguglio, B. Archer, J. Jafari and G. Wall (eds) London: Pinter.

Stewart, W. and Hull, R. (1996) 'Capturing the moments: Concerns of in situ leisure research', *Journal of Travel and Tourism Marketing*, 5(1/2), 3–20.

Stiglitz, J.E. (2002) *Globalisation and its Discontents*. New York: W.W. Norton.

Suarez Salazar, L. (1999) *Cuba: Isolation or Reinsertion in a Changed World?* Havana: Jose Marti.

Sunday Guardian, 6 September (1998), *Tourism Shakes Up Cuban Society*, 35.

Swarbrooke, J. (1999) *Sustainable Tourism Management*. Oxon: CABI Publishing.

Sylvester, C. (1999) 'Development Studies and Post-Colonial Studies: Disparate Tales of the Third World', *Third World Quarterly* 20, 4: 703-721.

Tearfund (2001) *Tourism: Putting Ethics into Practice*. Middlesex: Tearfund.

Thomas, N. (1992) 'The Inversion of Tradition', *American Ethnologist* 19, 2: 213-232.

Thurot, J.M. and G. Thurot (1983) 'The Ideology of Class and Tourism: Confronting the Discourse of Advertising', *Annals of Tourism Research* 10, 1: 173-189.

Tisdell, C. and C. Wilson. (2001) 'Wildlife-based Tourism and Increased Support for Nature Conservation Financially and Otherwise: Evidence from Sea Turtle Ecotourism at Mon Repos', *Tourism Economics* 7: 233-249.

Tsing, A. (2000) 'The Global Situation' *Cultural Anthropology* 15, 3: 327-60.

Turner, V. (1969, 1977) *The Ritual Process: Structure and Anti-Structure*. New York: Cornell University Press.

Turner, V. (1973) The Centre Out There: Pilgrim's Goal. *History of Religions* 12: 191-230.

Turner, V. (1974) *Dramas, fields and metaphors: Symbolic action in human society*. New York: Cornell University Press.

Turner, V. and Ash, J. (1975) *The Golden Hordes: International tourism and the pleasure periphery*. London: Constable.

Turner, V. and E. Turner (1978) *Image and Pilgrimage in Christian Culture: Anthropological Perspectives*. Oxford: Basil Blackwell.

United Nations (2000) *A Better World for All*. New York: United Nations.

United Nations (2001) *Human Development Report 2001*. http://www.undp.org/hdr2001/back.pdf. As posted on June 2003.

United Nations (2002) *UNFPA Country Programme Outline for Cuba (2003-2007)*. United Nations Development Programme.

United Nations (2003) *UNFPA Country Programme Outline for Cuba (2004-2007)*. United Nations Development Programme.

Uriarte, M. (2002) *Cuba – Social Policy at the Crossroads: Maintaining Priorities, Transforming Practice*. Boston: Oxfam America.

Urry, J. (1990) 'The consumption of "tourism"', *Sociology* 24: 23-35.

Urry, J. (1990, 1997) *The Tourist Gaze: Leisure and Travel in Contemporary Societies*. London: Sage.

Urry, J. (1994) 'Cultural Change and Contemporary Tourism', *Leisure Studies* 13: 233-238.

Urry, J. (1995) *Consuming Places*. London: Routledge.

Uvin, P. (2002) 'On High Moral Ground: The Incorporation of Human Rights by the Development Enterprise', *PRAXIS The Fletcher Journal of Development Studies*, Vol XVII. As posted: http://fltecher.tufts.edu/praxis/xvii/Uvin.pdf, December 2003.

Van Gennep, A. (1960) *The Rites of Passage*. University of Chicago Press: Chicago.

Van Tuijl, P. (2000) 'Entering the global dealing room: Reflections on a rights-based framework for NGOs in international development', *Third World Quarterly* 21, 4: 617-626.

Var, T., J. Ap and C. Van Doren (1994) 'Tourism and World Peace'. In *Global Tourism: The Next Decade*. W. Theobald (ed.) Boston: Butterworth-Heinnemann.

Wall, G. (1997a) 'Sustainable Tourism – Unsustainable Development'. In *Tourism Development and Growth*. S. Wahab and J. Pigram (eds) London: Routledge.

Wall, G. (1997b) 'Is Ecotourism Sustainable?', *Environmental Management* 21: 483-491.

Wallerstein, I. (1974) *The Modern World System: Capitalist Agriculture and the Origins of the European World Economy in the Sixteenth Century*. New York: Academic Press.

Weaver, D. (1991) 'Alternative to mass tourism in Dominica', *Annals of Tourism Research* 3: 414-32.

Weaver, D. (2005) 'Comprehensive and Minimalist Dimensions of Ecotourism', *Annals of Tourism Research* 32, 2: 439-455.

Weaver, D. and L. Lawton (1999) *Sustainable Tourism: A Critical Analysis*. Griffith: CRC Griffith University.

Weiler, B. (1991) 'Learning or Leisure? The Growth of Travel-Study Opportunities in Australia', *Australian Journal of Leisure and Recreation* 1, 1: 19-22.

White, N. and P. White (2004) 'Travel as Transition: Identity and Place', *Annals of Tourism Research* 31, 1: 200-218.

Wilk, R. (1996) *Economies and Cultures: Foundations of Economic Anthropology*. Westview: Boulder.

Wilkinson, S. (2008) 'Cuba's Tourism 'Boom': A curse or a blessing?', *Third World Quarterly* 29, 5: 979-993.

Wilson, D. (1981) 'Tourism as an Anthropological Subject – commentary on Nash article', *Current Anthropology* 22, 5: 461-481.

Wonders, N. and R. Michalowski. (2001) 'Bodies, Borders and Sex Tourism in a Globalized World: A Tale of Two Cities – Amsterdam and Havana', *Social Problems* 48, 4: 545-571.

Wood, D. (1993) 'Sustainable development in the Third World: Paradox or panacea?' *The Indian Geographical Journal* 68: 6-20.

World Bank (2000) *Human Development Report*. New York: Oxford University Press.

World Bank (2001) World Development Indicators. https://publicatons.worldbank. org/WDI. As posted on March 2003.

World Commission on Environment and Development (1987) *Our Common Future*. Oxford: Oxford University Press.

World Trade Organisation (1998) *Working Group in the Relationship between Trade and Investment* January 7.

Index

For Product Safety Concerns and Information please contact our
EU representative GPSR@taylorandfrancis.com Taylor & Francis
Verlag GmbH, Kaufingerstraße 24, 80331 München, Germany